A PRIMER OF ENVIRONMENTAL TOXICOLOGY

LEA & FEBIGER 1992 PHILADELPHIA

Lea & Febiger
200 Chester Field Parkway
Malvern, Pennsylvania 19355
U.S.A.
(215) 251-2230

Library of Congress Cataloging-in-Publication Data

Smith, Roger P. (Roger Powell), 1932–
A primer of environmental toxicology / Roger P. Smith.
p. cm.
Includes index.
ISBN 0-8121-1468-X
1. Pollution—Toxicology. I. Title.
RA565.S56 1992
615.9′02—dc20 91-25236
CIP

Reprints of chapters may be purchased from Lea & Febiger in quantities of 100 or more.

PRINTED IN THE UNITED STATES OF AMERICA

Print number: 5 4 3 2 1

PREFACE

Within the past decade an anonymous Associated Press staffer filed the following classic account of the problems that may be encountered when one tinkers with ecology and the environment. It was attributed to Dr. Lamont C. Cole of Cornell University who was describing a World Health Organization effort to control malaria-bearing mosquitoes in Borneo.

The insecticide, DDT, was sprayed to kill the mosquitoes, and as expected, it was highly effective. But, cockroaches in the same area were more resistant, and they were able to survive with high levels of DDT in their bodies. Cockroaches are a staple in the diet of native long-tailed lizards, or geckoes. As the geckoes fed on cockroaches, they accumulated the cockroach burden of DDT, which affected the nervous systems of the lizards to the point that they were much less agile. Cats considered the geckoes a great delicacy, and they were able to gorge themselves because the lizards became much easier to catch. The even more concentrated DDT in the geckoes began killing off the cats, and rats from the surrounding jungle began moving in bringing with them the threat of bubonic plague. The solution was to parachute fresh cats into the village, and they soon had the rat infestation under control. Then, the thatch roofs of the native huts began caving in because the geckoes ate not only cockroaches but caterpillars as well. The liberated caterpillar population feasted on the thatching enough to weaken it to the point of collapse.

This book contains a litany of similar disasters as seen in the only true vision, hindsight. One wonders if humans will ever be intelligent enough to avoid disrupting the dynamic, delicate, and interrelated processes that have existed on earth for eons of time. It is our inability to forsee interrelationships that has caused so much grief. Dr. James Lovelock, independent biomedical scientist, inventor, and chemist, in his book, *Gaia, A New Look at Life on Earth* (Oxford University Press, 1979, reprinted 1987), formulated the ultimate in interrelatedness. Gaia was the Greek goddess of earth, and the Gaia hypothesis, in one of its many modifications, holds that the living and the nonliving things of the earth together represent a self-regulating system that maintains optimal conditions for itself (homeostasis). More hardheaded scientists believe that the living things are not necessary and that the system can regulate itself through the immutable laws of chemistry and physics. We hope that the Gaia hypothesis will never be tested because we wish for mankind a life on earth eternal, but it is long past time to start doing a better job of housekeeping.

This book is the outgrowth of lectures given to students at the Vermont Law School and at Dartmouth Medical School, including Ph.D. candidates in the Department of Pharmacology and Toxicology and undergraduates at Dartmouth College. Contributions have been made by Drs. Dennis Devlin, Robert Gosselin, Paul Kaplita, and Bill Roebuck. Bill and I have taught together in many courses and for many years. This book has been shaped not only by his formal contributions but also by many of his ideas as conveyed informally.

Roger P. Smith

Hanover, NH

CONTRIBUTORS

DENNIS J. DEVLIN, PH.D.
Exxon Bio-Medical Sciences, Inc.
Mettlers Road
East Millestone, New Jersey

ROBERT E. GOSSELIN, PH.D.
Professor Emeritus of Pharmacology and Toxicology
Dartmouth Medical School
Hanover, New Hampshire

PAUL V. KAPLITA, PH.D.
Sterling Research Group
25 Great Valley Parkway
Malvern, Pennsylvania

BILL D. ROEBUCK, PH.D.
Professor of Toxicology
Dartmouth Medical School
Hanover, New Hampshire

CONTENTS

1	SOME HISTORIC INSIGHTS	1
2	ACUTE OR SINGLE DOSE TOXICITY	11
3	LIMITATIONS AND USES OF THE ED50 OR LD50	21
4	INHALATION TOXICITY AND THE CONCEPT OF CHRONICITY	35
5	SUBCHRONIC AND CHRONIC TOXICITY TESTING	49
6	ABSORPTION, DISTRIBUTION, AND EXCRETION OF CHEMICALS—Bill D. Roebuck, Ph.D.	59
7	BIOTRANSFORMATION	77
8	FOOD AND PESTICIDE LAWS: SETTING TOLERANCES	83
9	SURVEY OF PESTICIDES	97
10	THE RISE AND FALL OF DDT	113
11	CHLORINES, BENZENE RINGS, AND DIOXINS	127
12	HEAVY METALS—Bill D. Roebuck, Ph.D.	141
13	THE SULFUR CYCLE AND REDUCING-TYPE AIR POLLUTION	151
14	THE NITROGEN CYCLE AND OXIDIZING-TYPE AIR POLLUTION	171
15	THE DETERGENT CONTROVERSY—Robert E. Gosselin, Ph.D.	187

16 INDOOR AIR POLLUTION 209

17 FOOD ADDITIVES AND CHILDHOOD BEHAVIOR DISORDERS—Paul V. Kaplita, Ph.D. 219

18 HAZARDOUS WASTE SITES—Dennis J. Devlin, Ph.D. 233

19 THE ALCOHOLS 255

20 CARCINOGENICITY, MUTAGENICITY, AND TERATOGENICITY 273

INDEX 289

1

CHAPTER

SOME HISTORIC INSIGHTS

Toxicology was probably among the earliest of medical sciences. Even before recorded history primitive humans must have known that certain plants were to be avoided as food because they produced unpleasant toxic effects or even death. If legend is correct, Cleopatra was well aware of the lethal effects of venomous animals. That slowly growing body of knowledge was passed down from generation to generation over time because it was of inherent practical value. Even in Cleopatra's time most people would take precautions to avoid being bitten by an asp. Over the centuries that same body of knowledge began to turn in a new and more sinister, but still highly pragmatic, direction. Probably by trial and error it was learned that the poisonous ingredients in certain plants could be applied to arrows, darts, or spears and used for hunting or warfare with much greater efficiency than projectiles that were not so treated. Some ancient societies used extracts from poisonous plants as instruments for the execution of undesirable members of their group.

Yet a third, and perhaps more socially commendable direction, was taken when it was learned over the centuries that some poisonous plants, when taken in proper amounts or doses, were actually beneficial in certain diseases, giving rise to the scientific disciplines of pharmacy, pharmacology, and medicine. Because of several unique circumstances, the oldest records of poisons and useful remedies came to us from the Egyptians of about 1500 BC. The Egyptians invented a primitive form of paper, called papyrus, which was made from a plant of the same name. They also devised an advanced symbolic language in the form of hieroglyphics. Finally, they had a propensity for sealing written records into tombs that would preserve them intact and still legible for thousands of years awaiting discovery in relatively modern times. The so-called Eber's papyrus listed more than 700 remedies and poisons together with directions for collecting the natural materials and for preparing and administering them. Most of these may be instantly recognized today as pure quackery, but a handful still occupy important niches in modern medicine. They knew about the usefulness of a drug that has come to be called colchicine for the painful symptoms of an acute attack of gout, although they used it as a crude extract from the plant, meadow saffron. The use of morphine as contained in opium was known to relieve other types of pain, and they also recognized the possibly fatal consequences of an overdose of the same material.

The ancient Greeks and Romans were diligent practioners of the art of "applied toxicology." Hemlock, which was responsible for the death of Socrates, was the official state poison for Greece. During the Middle Ages, the Borgia family was notorious for their alleged use of arsenic in wine. Cesare Borgia, his half-sister Lucretia, and their father Pope Alexander VI, became wealthy and powerful through the misfortunes they inflicted on others. Some say that Lucretia was actually innocent of any wrongdoing, but her name is still firmly linked to the surreptitious use of poison. It is also said that the Borgia dynasty came to an end with a poetic stroke of justice when an ill-informed servant served the Pope the wrong bottle of wine. There is a theory that lead poisoning of epidemic proportions contributed to the downfall of the Roman Empire. It was then the practice to boil down grape juice in lead pots into syrups that were used to color, flavor, or preserve other foods and beverages. During this process the syrups inevitably became contaminated with lead from the pots.

During the Renaissance, an enlightened figure with the unlikely name of Phillipus Aureolus Theophrastus Bombastis von Hohenheim Paracelsus (small wonder that he used the pen name "Paracelsus" instead of "Bombastis") formulated some principles of toxicology that are still relevant today. Paracelsus was a strong advocate of the experimental approach for studying the toxic effects of natural substances. He recognized that at least sometimes the therapeutically useful and the toxic effects of these materials were expressions of different degrees of intensity of the same basic activity. Perhaps this, the most important of his principles, could be stated as: The effect is determined by the dose, from which it follows that all chemicals are potentially poisonous.

In 1713, a remarkable Italian physician, Bernardino Ramazzini, published a book called (in translation) *Diseases of Workers.* That volume was to lead to his recognition as the father of occupational medicine. Many of the maladies that he described were caused by occupational exposure to toxic chemicals. A Spanish physician, Mattieu Joseph Bonaventura Orfilia (1787–1853) is generally credited with defining the science of toxicology and for providing the impetus for its emergence as a separate and distinct discipline.

In the 19th century, Paul Ehrlich (no relation to *The Population Bomb* author) showed the scientific world how specificity with respect to therapeutic or toxic effects could be conferred on a chemical by making modifications in its structure. Starting with the instrument of the Borgias, arsenic, which was known then only as a deadly poison, he patiently threaded his way through 604 different chemical forms before he discovered E–(for Ehrlich, of course) 605. This 605th derivative was the first truly effective drug against syphilis that was safe enough to be given to humans.

The felt hat industry has been traced to the mid seventeenth century in France, and it was probably introduced into England some time around 1830. In Turkey, camel hair was used, and it was discovered that the felting process was speeded up if the fibers were moistened with camel urine. It is said that in France workmen used their own urine, but one particular workman seemed consistently to produce a superior felt. This person was being treated with a mercury compound for a venereal disease, and an association was made between mercury treatment of the fibers and an improved felt. Eventually the use of solutions of mercuric nitrate was widespread in the felt industry, and mercury poisoning became endemic. The symptoms included lassitude, headache, insomnia, weakness, muscle pains, weight loss, tremor, and sometimes frank psychosis. On December 1, 1941, the United States Public Health Service (USPHS) banned the use of mercury in the felt hat industry in this country. Although it has been suggested that the expression "mad as a hatter" and the character portrayed in Lewis Carroll's *Alice*'s *Adventures in Wonderland* may have had other origins, few can resist making this apocryphal analogy (Goldwater, 1972).

The discipline of toxicology may be subdivided in many different ways. One way is on the basis of the kinds of experiments that toxicologists perform. One set of toxicological experiments is designed to evaluate how poisonous or toxic a given chemical is in order to define the safe level of exposure. This is quantitative toxicology, or as it is sometimes called, safety evaluations. From a societal standpoint this may well be the most important subarea of toxicology. The other kinds of experiments are directed more qualitatively at how a chemical exerts its toxic effects in the body. This subarea is called mechanistic toxicology. Sometimes these kinds of experiments make fundamental contributions to our store of knowledge about how living things carry out their normal functions. Poisons were used with dazzling virtuosity for such purposes by a nineteenth century French physician and physiologist named Claude Bernard who authored *An Introduction to the Study of Experimental Medicine*. Bernard was unshakable in his conviction that all of life's processes obeyed natural laws of chemistry and physics, and that these processes could be studied rationally by designing experiments that led to perturbations in the natural process. The introduction of a toxic chemical into biological systems was one way to produce such perturbations. That such an approach can yield fruitful results is illustrated by his experimental work with carbon monoxide (CO).

As early as 1846, CO was reported to be poisonous, and it was recognized as a potential environmental problem arising from the combustion of fossil fuels. Bernard had no preconceived notion about how or why CO might be toxic when he began his studies. His first experiment was the essence of simplicity. He exposed a dog to CO until it died (proving that, indeed, it was toxic), and then he performed an autopsy. Immediately he made an observation that suggested a hypothesis. Under ordinary circumstances the blood in the left side of the heart and in the arteries leading away from the heart is bright red. That is because these areas contain blood that has just come from the lungs, and it is saturated with oxygen (O_2), which is carried in combined form with hemoglobin in red blood cells. The function of the blood is to transport its load of O_2 to the rest of the body. As the O_2 is extracted from blood by the various organs and tissues and used by them, the blood changes color and becomes more purplish accounting for the color of veins as opposed to arteries. It returns from its journey to the most distant parts of the body to the right side of the heart from which it is pumped to the lungs for resaturation with O_2. In the case of the dog poisoned with CO, however, the blood was the same bright red color in veins as in arteries and in the right side of the heart as in the left.

From that observation Bernard's first hypothesis was that because the blood in veins was still a bright red color somehow the body had failed to extract its content of O_2. If this hypothesis was correct, then CO must be doing something to prevent the extraction of O_2 from blood or to prevent its utilization by various organs and tissues. This hypothesis suggests another experiment because it follows that the venous blood in CO-poisoned animals should contain the same amount of O_2 as the arterial blood. When Bernard actually measured the O_2 content of both venous and arterial blood, however, he found to his surprise that neither contained much O_2.

Obviously, the first hypothesis was incorrect, and it was now necessary to formulate a new one that incorporated the newly established experimental fact that CO seemed to prevent the normal oxygenation of blood even though it was the appropriate color for arterial blood. Even in those days it was known that gases could displace each other. So, Bernard's second hypothesis was that CO had displaced O_2 and driven it out of the blood. The new hypothesis called for a new experiment. Blood was taken from a normal healthy animal and exposed to air until it was saturated with O_2. This blood sample was then placed in a sealed vessel containing CO and shaken vigorously for a time. The gas over the blood was then analyzed chemically. It was found that the CO content of the gas had decreased, and that it had been replaced by an equal volume of O_2. Bernard, thus, reached the conclusion that indeed CO had displaced O_2 from its binding sites on hemoglobin. In doing so, the CO remained bound to hemoglobin creating a new chemical form of hemoglobin, carboxyhemoglobin, that had a reddish color similar to that of O_2-saturated blood. From this it follows that CO is poisonous because it prevents the blood from performing its vital function of transporting O_2.

The brain uses large amounts of O_2, and, therefore, is the most sensitive organ in the body to O_2 deprivation. The end result of CO poisoning is a functional asphyxia, which is analogous in many respects to O_2 deprivation caused by physical obstruction, such as strangulation or drowning. The same line of reasoning suggested the basis for the modern treatment of CO poisoning after exposure has been terminated. Because the signs and symptoms of CO poisoning are largely, if not exclusively, caused by a lack of O_2, one need only furnish the victim with O_2 in large amounts, and the O_2 will assume the role of a specific "antidote." Instead of providing air, which contains only 20% O_2, the victim should be made to breathe pure (100%) O_2 from a pressurized tank. In severe cases and if a hyperbaric chamber is available, pure O_2 can be provided at pressures that are more than double the usual 1 atmosphere to further accelerate recovery.

In 1845, Bernard was given a sample of the South American arrow poison, curare. Curare had been used for centuries by natives of the Amazon basin for hunting game. At that time nothing was known about its mechanism of action except that poisoned animals seemed to be paralyzed. Death occurred rapidly if the poison was injected into the blood stream, but the flesh from poisoned animals could be eaten with impunity. Obviously, the route by which the poison gained access to the body was critical for its toxic effects. When Bernard injected curare into a frog, it died within a few minutes. He then performed an autopsy, and found that the heart was still beating normally. The hearts of many species, including humans, will continue to beat for variable periods of time even after complete removal from the body. The blood had a normal appearance, and there was no obvious damage to any critical organ. In normal frogs electrical stimulation of nerves leading to skeletal muscles will cause the muscles to contract for at least a few hours after death. Bernard made the critical observation that such was not the case with his curare-poisoned frog, which substantiated the reports of paralysis in poisoned game.

Contraction of skeletal muscles involves a complex relay of electrical and chemical signals that may originate in the brain, and travel down the spinal cord to nerves that lead out to the muscles. Other nerves transmit messages from the muscles back to the spinal cord to complete a reflex arc, which may not involve the brain. Bernard next discovered that curare was still effective in frogs in which the brain had been surgically removed, and this observation eliminated the brain as a site of action for curare. The remaining sites where curare might act included the spinal cord, the nerves to muscles, the muscles, or the narrow junctional gap between the nerve terminus and the muscle, the so-called synaptic cleft.

In his next experiment, Bernard isolated and surgically freed a nerve fiber leading to a muscle in the hindlimb of a frog. He then tied off that same leg with a suture to totally obstruct blood flow into the area. Next curare was injected into the body of the frog. It produced paralysis to electrical stimulation of nerves leading to all the muscles that he tested except those in the tied-off leg. His conclusion was that the obstruction to blood flow in that leg had prevented curare from reaching its site of action. When he stimulated the nerve leading away from the muscle in that leg back to the spinal cord and then out again to the muscle (reflex arc), the muscle in the tied-off leg again contracted. Both of these observations eliminated the spinal cord as the site of action of curare.

The isolated nerve leading to the muscle in the tied-off limb was then soaked in curare. Electrical stimulation of that nerve still produced muscle contraction, which eliminated the nerve as the site of action. When he electrically stimulated muscles directly in the body of the frog, they contracted even though stimulation of the nerves going to the same muscles failed to produce such an effect. Thus, curare was not acting on the muscle. By exclusion Bernard had limited the site of action of curare to the area between the nerve and the muscle, or the synaptic cleft.

It is now known that when an electrical stimulus is applied to a nerve it initiates a wave of electrical activity (an action potential), which travels the length of the nerve in both directions (conduction) unless one is able to stimulate precisely at one end or the other. What happens when the action potential reaches the synaptic cleft (transmission) was a subject of controversy in Bernard's time and for some time thereafter. One hypothesis was that the electricity jumped the gap to the muscle or the next nerve (the "sparks" school). The alternative hypothesis was that the electrical activity stopped at the nerve terminus, but before it did, it caused a chemical to be released from the nerve ending, which floated across the synaptic cleft and excited a specialized area on the muscle (a receptor), which initiated a new action potential in the muscle or the next nerve (the "soup" school).

The soup school is now recognized as having had the correct hypothesis. What remained to be done was to separate isolated nerves from their muscles and stimulate the nerve in a separate container containing a small amount of solution. The nerve could then be discarded, and the solution applied directly to the muscle, which then contracted. The chemical that was released from the nerve ending is now called acetylcholine, and it is only one example of a wide variety of naturally occurring chemicals called neurotransmitters. Acetylcholine excites only the minute specialized areas called receptors on the skeletal muscle (or nerve) membrane, but the excited receptor is able to initiate a chemical cascade that sends an action potential down the muscle membrane to cause contraction or conduction down the next nerve membrane. Curare acts by sitting on those receptors so that acetylcholine cannot get to them. If this excitatory neurotransmitter is denied access to its receptors, the result is muscle paralysis. A purified form of curare is now widely used in human patients for promoting muscle relaxation during surgery, but the experimental work was an intellectual milestone toward proving the validity of the neurotransmitter hypothesis.

It is difficult today to try to place Bernard's achievements in an appropriate perspective. He cannot be compared with his peers because in his chosen field there were none. Only a handful of experimentalists in the life sciences were alive at the time, and most of them still believed that at least some natural phenomena could not be fathomed by humans; they occurred as miracles in witness to a supreme intelligence. In contrast, Bernard was totally dedicated to his conviction that all of life's processes were governed by natural laws. He was so productive and brilliant in the laboratory that in the brief span of twenty years, which encompassed his entire active career, he discovered and deduced more dominating facts in medicine and physiology than all of the other experimentalists in the world up to that time combined. It was remarked on the occasion of his death that "he was not merely a physiologist, he was Physiology." Never again could divine intervention be offered as an explanation for the unknown, nor could the unknown be equated with the unknowable.

SUMMARY POINTS

All chemicals are toxic; only the dose accounts for the difference between a toxic and a beneficial effect.

The toxicity of a chemical can be modified radically by making changes in its chemical structure.

One preoccupation of toxicologists is to make estimates about the potential hazards of given chemicals (also called risk assessment).

The other major preoccupation is to find out how chemicals exert their characteristic toxic effects or what the chemical does to the body.

Carbon monoxide combines with hemoglobin in red blood cells to the exclusion of oxygen producing an internal suffocation.

Curare blocks receptors on skeletal muscle to block normal access of the neurotransmitter, acetylcholine, resulting in paralysis and suffocation.

BIBLIOGRAPHY

Adams E: Poisons. Sci Am 201: 76, 1959.

Bernard C: An Introduction to the Study of Experimental Medicine (translated by HC Greene). New York, Macmillan, 1927.

Goldwater, LJ: Mercury, A History of Quicksilver. Baltimore, York Press, 1972.

Klaassen C, Amdur M, Doull J: Cassarett and Doull's Toxicology, The Basic Science of Poisons, 4th ed. New York, Macmillan, 1991.

Levine RR: Pharmacology: Drug Actions and Reactions, 3rd ed. Boston, Little, Brown and Co., 1983.

Meek WJ: The gentle art of poisoning. JAMA 158:335, 1955.

Ramazzini B: Diseases of Workers, (translated by WC Wright). New York, Hafner, 1964.

2

CHAPTER

ACUTE OR SINGLE DOSE TOXICITY

The terms "poisonous" and "toxic" are synonomous, but these are not necessarily synonomous with "hazardous." Being toxic is only one of the ways in which a chemical may be hazardous. Alkaline drain cleaners are hazardous because they are corrosive, gasoline because it is flammable, compressed air only because it is under high pressure, and radon because it is radioactive. Toxic chemicals are thought of as acting from within the body, that is, they are said to act systemically. In order to act systemically they must gain access to the interior of the body, i.e., they must be absorbed. Once they are absorbed, they have more or less specific effects in that they may act on a particular organ, block the actions of a specific enzyme or occupy a receptor. (See the discussion of curare in Chapter 1). In the case of CO (see Chapter 1) the ultimate target organ was the brain, but it was affected indirectly through an action of CO on the blood. The CO was absorbed into the blood through the lungs.

The above distinctions between toxic and hazardous, however, are not mutually exclusive and, therefore, they are somewhat ambiguous. For example, with respect to explosive hazard, the old nursery rhyme tells us about Little Willie whose curiosity was aroused when he found some dynamite, with disasterous consequences. Willie, however, might also have indulged his curiosity by chewing on a stick. Dynamite contains highly explosive nitroglycerin in a stabilized form. Nitroglycerin is also useful in small doses to relieve the pain in the heart caused by an attack of angina. By dilating the coronary arteries and blood vessels elsewhere in the body, blood flow to the heart is improved and its workload is reduced. The pain that arises from an inadequate supply of O_2 to overworked heart muscle is thereby relieved. An overdose of nitroglycerin, however, in accord with the Paracelsus Principle that everything is a poison, may dilate blood vessels excessively to the point where the supply of blood O_2 to the brain is inadequate, which may result in fainting. Thus, dynamite is also capable of producing systemic poisoning. This sort of ambiguity applies to many of the examples of hazardous chemicals cited above.

As already noted, toxicologists have two fundamental goals. The first, to elucidate how poisons work in the body, was illustrated in Chapter 1 with the experiments of Claude Bernard on CO and curare. The second goal is to define in quantitative terms the relationship between the dose or concentration of a chemical and the biological response to it. The concept that a definable dose-response relationship exists is a keystone to toxicology and its related disciplines. As a general rule, the response tends to appear more quickly and to intensify as the dose is increased. Allergic reactions, however, are a recognized exception in which the response intensity is not influenced greatly by increases in dose. Moreover, allergic reactions never occur on the first exposure to a chemical. At least one "sensitizing" exposure is required before allergy can develop. In contrast, a toxic chemical can produce poisoning with the first dose. A pessimistic toxicologist is concerned about the toxic range of doses, whereas the optimistic toxicologist wants to define the safe range of doses or exposures. The facts are such that neither can do a complete job without the other.

With the possible exception of carcinogens (see Chapter 20), there exists for most chemicals a range of doses below a certain critical threshold for which no observable toxic effect occurs, the no observable effect level or NOEL. Above the NOEL, one enters the toxic range of doses. Here the response intensifies and sometimes changes qualitatively with increasing doses all the way up to the ultimate toxic effect, death. In the absence of artificial life support, death is a sharply defined endpoint. If the goal is to define quantitatively the dose-effect relationship for a toxic chemical, it follows that one must give doses that are in the toxic range. Below that range, by definition, there is no effect to observe or quantify.

The major reason for testing chemicals for their toxicity in laboratory animals is for the purpose of extrapolating the results to humans. Unfortunately, animals are imperfect models for human toxicity. Indeed, any one species may exhibit responses to chemicals that differ quantitatively or even qualitatively from that in another species. One of the most obvious differences between species is size or body weight. A rat may weigh 250 g, whereas a human may weigh 70,000 g, a 280-fold difference. Obviously, one cannot give the same total amount of chemical to each and expect to make any meaningful comparison. Thus, doses are expressed in terms of units of body weight, e.g., milligrams of chemical A per kilogram of body weight or mg/kg. On rare occasions in human medicine when great precision is required the dose may be expressed in terms of units of body surface area, e.g., grams/square meter or g/m^2.

A greater degree of sophistication, particularly when making comparisons between or among chemicals, is achieved by expressing doses in terms of molecular weights. Because the chemicals of interest may have widely different molecular weights, this convention relates the dose to the number of molecules of the chemical given. Atomic weights are the relative weights of all the elements in the periodic table compared to a standard atom. The standard atom used is a form of carbon with an atomic weight of 12. Molecular weights are the sum of all the atomic weights in a given molecule. When the molecular weight is expressed in grams it is called a gram molecular weight or mole (mol). One mole of chemical A contains the same number of molecules as one mole of any other chemical. Because gram molecular weights tend to be large numbers in terms of the range of toxic doses of many chemicals, it is more reasonable to use millimoles (mmol or mol/1000). To convert a dose of 10 mg/kg of sodium cyanide, which is lethal in some species, to mmoles per kilogram, divide the dose by the molecular weight expressed in appropriate units:

$$(10 \text{ mg/kg})/(50 \text{ mg/mmol}) = 0.5 \text{ mmol/kg}$$

It is important to remember that toxicity is inversely related to dose: toxicity = 1/dose. The smaller the dose required to produce the response, the more toxic is the chemical.

Table 2–1 shows how expression of the dose in terms of millimoles per kilogram avoids ambiguities about which chemical is more toxic. On the basis of absolute weights (mg/kg) as shown in the footnote to Table 2–1, chemical A, irrespective of its identity or molecular weight, would always be said to be more toxic than chemical B. However, as the chemical identities and molecular weights of A and B are varied, one might conclude correctly that A was more toxic, that B was more toxic, or that A and B were equitoxic. When substituting the molecular weight convention for the absolute weight convention, large differences in alleged toxicity may disappear. On other occasions two chemicals said to be equitoxic on the basis of absolute weights may be shown to have very different toxicities on the basis of their molecular weights.

TABLE 2–1. ADVANTAGE OF EXPRESSING DOSES IN TERMS OF MOLECULAR WEIGHTS RATHER THAN ABSOLUTE WEIGHTS*

Chemicals and Molecular Weights	Toxic Dose of A (mmol/kg)	Toxic Dose of B (mmol/kg)	More Toxic Chemical
A > B (1000, 100)	0.01	1.0	A
A = B (1000, 1000)	0.01	0.1	A
A < B (100, 1000)	0.1	0.1	A = B
A < B (100, 10,000)	0.1	0.01	B

*In this table the toxic dose of A is always taken to be 10 mg/kg, and the toxic dose of B is always 100 mg/kg irrespective of the chemical identities and molecular weights of either A or B. As the identity and molecular weights are varied in the left column, the conclusion about which chemical is more toxic also changes.

In going from the dose to the response side of the relationship, one finds that two basic kinds of biologic responses are recognized. The all-or-none (quantal) response is like a light switch that has only two positions, "on" or "off." Thus, the response either occurs or it does not occur. For example, after a given dose of a chemical a laboratory animal may either live or die, alternatively, it may remain awake or go to sleep. Death is the least ambiguous of these two quantal responses because in the second example the fact that sleep can vary in intensity or duration has been ignored. By taking these latter parameters into consideration and changing the nature of our observation to that of the duration of sleep, we can convert our quantal response into a graded or variable intensity response. Other examples include decreases in body temperature or weight with increases in dose. When a quantal or graded response is evaluated after a single dose of a chemical, the result is called an acute effect.

Many years ago, it was common to express acute toxicity in terms of the lowest dose that had ever been observed to produce the effect. For example, the lowest dose of a chemical that had ever been observed to result in death was called the minimal lethal dose (MLD). Philosophically it seemed that establishing the lowest dose of a chemical that might be very dangerous would be extremely relevant to toxicology. In practice, it was recognized that animals often survived even when given doses larger than the MLD. Moreover, the most recently established value for a given MLD was always smaller than the last established value in that species. It was recognized that both of these phenomena resulted from animal-to-animal variation in sensitivity to the toxic effects of the chemical. In any population of laboratory animals a small fraction is sensitive to the chemical and responds to very low doses. Another small fraction of equal size is resistant and requires very high doses to produce the same effect. As larger and larger numbers of animals are tested, the chances of encountering animals at one or the other extreme of the population increase. Thus, the concept of biologic variability was implicitly accepted by early toxicologists, but it was ignored operationally because the tools for dealing with it in quantitative terms had not yet been devised.

The distribution of sensitivities to a toxic effect of a chemical in a population is represented in Figure 2–1. Most of the population clusters around the average or mean toxic dose. The more sensitive fraction of the population lies to the left of the mean toxic dose, and the more resistant fraction of the population lies to the right of the mean toxic dose. If death is the effect being monitored, one can derive a parameter called the mean lethal dose. Unfortunately, this parameter can be, and often has been, abbreviated as "MLD" resulting in confusion with the minimal lethal dose. Because it is the derived or calculated dose that is lethal to

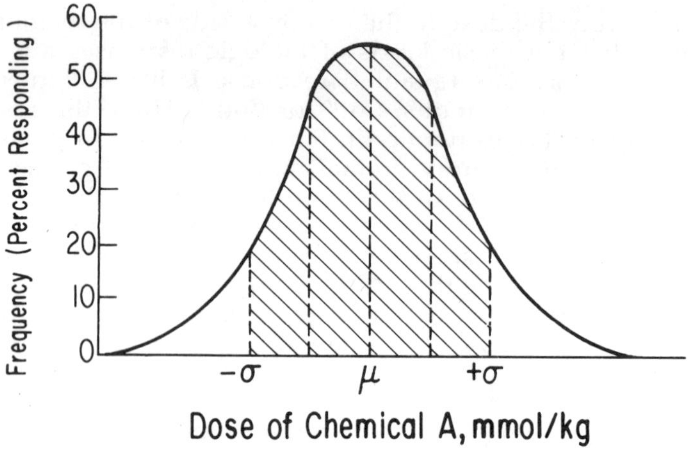

Figure 2–1. The distribution of sensitivities to a toxic effect of chemical A in a population of laboratory animals where μ = the mean toxic dose and $-\sigma$ and $+\sigma$ indicate toxic doses at one standard deviation (μ) on either side of the mean. (After Loomis TR: Essentials of Toxicology, 3rd ed. Philadelphia, Lea & Febiger, 1978.)

50% of a given population, it is abbreviated less ambiguously as the LD50.

The bell-shaped distribution of sensitivities, illustrated in Figure 2–1 is common, at least in approximate form, to a great many biological variables that can be measured, e.g., the distribution of body weights in a population or the heights for a given age. It is called a normal or Gaussian distribution. As a method for graphing the distribution of sensitivities to a toxic effect of a chemical in a population, however, it is not satisfactory. In Figure 2–1 the dose represented by $-\sigma$ extrapolates to about 16% of the population responding. However, the dose represented by $+\sigma$ also extrapolates to about 16% responding. Because the percentage of the population responding must increase with increasing dose, what is needed to express the results in a more appropriate way is a cumulative form of the normal distribution. If the area under the curve in Figure 2–1 is progressively summed up from left to right and expressed as a percentage of the total area, the integral or cumulative form of the normal distribution is obtained as seen in Figure 2–2. Now the percentage of the population responding to doses above the LD50 can be extrapolated directly. The use of the logarithmic scale on the abscissa has several advantages: (1) Often a wide range of doses is needed to define the entire dose-response relationship and the logarithmic scale conserves space. (2) The logarithmic scale tends to make the curve

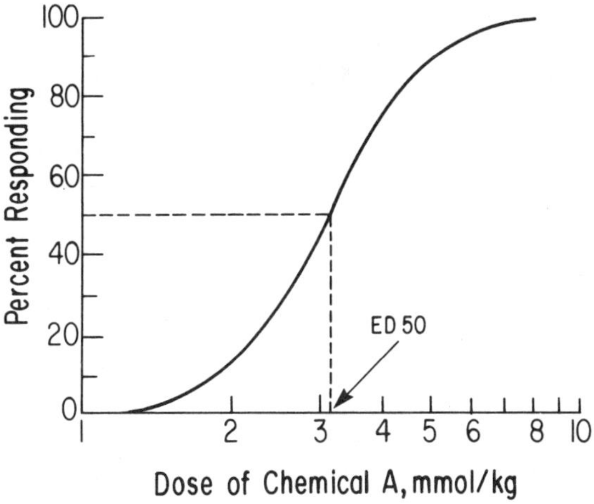

Figure 2–2. The cumulative (integral) dose-response relationship for a toxic effect of chemical A in a population of laboratory animals. The area under the curve in Figure 2–1 has been summed up from left to right and expressed as a percentage of the total area. (After Goldstein A, Aronow L, Kalman SM: Principles of Drug Action: The Basis of Pharmacology. New York, Harper & Row, 1968.)

more linear around its middle portion, the region of greatest interest because it includes the effective dose 50 (ED50) or the LD50. And (3) use of the logarithmic scale allows one to plot doses directly instead of converting them to logarithmic units.

Any normal distribution is completely defined by only two parameters, the mean or average value in the distribution (μ) and the degree of spread of the distribution, which is a reflection of the variability of the individual measurements. The latter is called the standard deviation (σ). Because both of these parameters vary depending on the measurements being made and other factors, an infinite number of normal distributions exist. Special interest, however, has centered around one particular normal distribution in which $\mu = 0$ and $\sigma = 1$. There are great advantages to rescaling by graphic or mathematic means any other normal distribution to that particular configuration. Extensive tables and powerful statistical tests have been developed for that particular distribution. For example, in a distribution with $\mu = 0$ and $\sigma = 1$, 68% of the population lies within $\pm 1\sigma$ of μ (shaded area in Figure 2–1) and 95% of the population lies within $\pm 2\sigma$ of μ. Any set of dose-response data that shows a normal distribution can be rescaled to this special normal distribution. Any measurement taken from a population with a mean of μ and standard deviation of σ can be rescaled by using the equation:

$$z = (x - \mu)/\sigma$$

The parameter, z, which is sometimes called a normal equivalent deviant, allows one to use the extensive statistical tables for the normal distribution.

A graphic way of "normalizing" dose response data is shown in Figure 2–3, which is a representation of probit graph paper. This kind of paper has been used for many years by toxicologists and pharmacologists to derive ED50 and LD50 values. A dose-response relationship and its ED50 have been plotted in Figure 2–3. The bell-shaped curve and its accompanying numbers and symbols on the ordinate do not appear on probit paper as it is purchased. Values of 0 and 1 have been assigned to μ and σ, respectively, to obtain normal equivalent deviants. By adding 5 to each of these, new numbers are obtained, which are called probits, and they all have positive values. The secret to probit

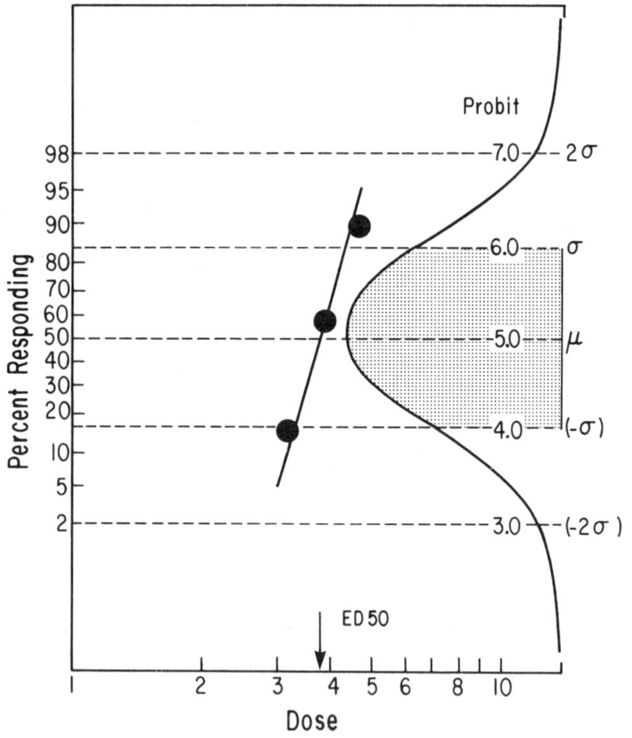

Figure 2–3. The dose-response relationship for a toxic effect of chemical A in a population of laboratory animals as plotted on probit graph paper. The curve drawn on the ordinate is intended to show how the spacing on the right ordinate has been adjusted to conform to a normal distribution with mean (μ) = 0 and standard deviation (σ) = 1.

graph paper lies in the unusual spacing on the ordinate. The spacing is very close together between 16 and 84% of the responding population, i.e., $\pm 1\sigma$ about μ. It is much more spread out between the values below 2.5% of the population and above 97.5% of the population, which includes $\pm 2\sigma$ about μ. Thus, the spacing reflects the distribution of expected values for a normal distribution with $\mu = 0$ and $\sigma = 1$.

Probit graph paper is almost as obsolete in modern toxicology as the slide rule, but it is still useful for first approximations of the ED50 or LD50. It also illustrates how one actually goes about determining an ED50. Groups of animals are used, and all the animals in a given group receive the same dose of the chemical. The number of animals in each group is not critical. Experience has shown, however, that it is difficult to obtain meaningful results with groups smaller than 5, and that satisfactory results can be obtained with 10 or more animals in a group. With a previously untested chemical and any clearly measurable endpoint, one starts with a small group and gives a dose that is only a guess. If none of the animals respond, the dose is increased for the next group. If all of the animals in the group respond, the dose is lowered for the next group. The objective, as shown in Figure 2–3, is to arrive at at least three doses for which some of the animals in each group respond to plot the percentage responding on the ordinate. This criterion eliminates responses of both 0 and 100%, but such responses are not meaningful in any event. With a 0% response one cannot be sure how far below the critical threshold for a response the dose actually was, and with a 100% response one cannot be sure how far above the minimal dose for a 100% response the dose given actually was. An adequate experiment would have at least 3 data points lying within $\pm 1\sigma$ of μ, and which lie approximately on a straight line. Doses are sometimes spaced 0.5 logarithmic units apart, but such spacing is not always appropriate. When the data have been plotted and the curve drawn to the best fit by eye, the ED50 can be extrapolated directly on the abscissa.

The modern digital computer and sophisticated software packages have nearly eliminated the need for probit graph paper. A good modern computer program will compute the best fit straight line for the data, the LD50, and its 95% confidence limits. The latter are extremely useful. They indicate that if one did the same experiment over again 100 times, the mean ED50 would fall outside of those upper and lower limits because of chance alone only 5 times. Thus, the probability of any one derived ED50 falling outside of the limits because of chance only is said to be less than or equal to 5% ($p < 0.05$). The computer can be programmed to provide the slope of the dose-response curve and its 95% confidence limits. Finally, it can provide a parameter called the potency ratio or the ratio of two ED50 values for two different chemicals or for the same chemical given under two different conditions. Potency ratios and their 95% confidence limits are also useful in testing for interactions between chemicals or for comparing the toxicities of two chemicals (see Chapter 3).

SUMMARY POINTS

Chemicals must be absorbed in order to exert systemic toxic effects.

For most chemicals there is a threshold dose, the NOEL, below which no toxic effects can be observed.

A definable relationship exists between doses above the NOEL and the biologic response that is observed.

The response may be quantal (all-or-none) or graded (variable intensity).

The sensitivities to various doses are normally distributed in animal and human populations.

The sensitivity of a population can be expressed in terms of a calculated or extrapolated dose to which 50% of the individuals tested respond.

BIBLIOGRAPHY

Goldstein A, Aronow L, Kalman SM: Principles of Drug Action: The Basis of Pharmacology. New York, Harper & Row, 1968.

Loomis TA: Essentials of Toxicology, 3rd ed. Philadelphia, Lea & Febiger, 1978.

Snedecor GW, Cochran WG: Statistical Methods, 6th ed. Ames, IO, The Iowa State University Press, 1967.

Spratt, JL: Computer program for probit analysis. Toxicol Appl Pharmacol. 8:110, 1966.

3

CHAPTER

LIMITATIONS AND USES OF THE ED50 OR LD50

It is likely that as a toxicologic test required by government agencies for regulatory purposes, the LD50 may soon be obsolete. Concern has been expressed from many quarters about the large numbers of animals that are sometimes used to determine a parameter that in and of itself has many limitations. There are, however, many ways of reducing the numbers of animals specified in the "official" test without losing precision or accuracy. The number, however, cannot be reduced to zero because the need for the test in legitimate scientific inquiry, such as the search for new antidotes to toxic chemicals, will certainly continue, and there will always be a need by regulatory agencies for an approximation of an LD50.

One of the limitations of an LD50 has been alluded to already, namely, differences among species. Although prominent differences are not always seen, they are so common that they should always be anticipated. Indeed, one entire subarea of toxicology, the pesticide industry, commercially exploits species differences in toxicity (see Chapter 9). There is a continual search for chemicals that are selectively toxic to insect, plant, or animal pests, and that are relatively nontoxic to humans or nontarget wildlife. Two examples of extreme species differences in toxicity follow. Sodium fluoroacetate (Compound 1080) is a rat, mouse, and coyote poison so dangerous that it can only be used by licensed applicators. Its oral LD50 in the dog is 0.06 mg/kg, whereas in the South African clawed toad it is greater than 500 mg/kg, a species difference of over 8000-fold. An example involving two species that are much more closely related phylogenetically is the dioxin contaminent in Agent Orange, TCDD (see Chapter 11), where the acute oral LD50 in guinea pigs is 1 microgram/kg. (A microgram (μg) is 1 one-millionth of a gram or 1 one-thousandth of a milligram.) In hamsters the comparable value is 5000 μg/kg. A common estimate for the average species difference in toxicity for all types of chemicals is 10-fold.

Even within a given species, various strains have different genetic or immunologic makeups that may give rise to differences in sensitivity. For example, various strains of rats have been identified and inbred specifically because of their greater susceptibility to some forms of chemically induced cancer. Using such strains in testing for the carcinogenic effects of chemicals is believed to decrease the chances that such effects will be overlooked. Conversely, the use of such strains probably increases the chances for false positive findings with respect to other species and humans.

Less commonly, sex may be an important determinant of toxicity. A striking sex difference recognized since 1945 is seen with respect to chloroform-induced kidney damage in mice. Chloroform is a common laboratory chemical, and in a number of institutions it was noted that when mice were accidentally exposed to low concentrations, the males of some strains would die within a few days, whereas the females similarly exposed would remain healthy. On the other hand, benzene is widely held to be more toxic to the human female than to the male. Some years ago expectant mothers were sometimes given a synthetic female sex hormone (diethylstilbesterol, DES) during early pregnancy if they had a history of spontaneous abortion. Years later this practice was associated with an increased incidence of a rare form of vaginal cancer in the female offspring when they reached puberty.

Age and body weight generally parallel each other in growing laboratory animals so that it is not always possible to know whether one is a more critical determinant of toxicity than the other. In a few special instances each has been shown to be more important. Human toddlers are well known to be more sensitive to the dangerous toxicity of aspirin than older children or adults even when appropriate corrections are made for the differences in body weight. One cannot generalize, however, that the young are always more sensitive to toxic chemicals than older members of the same species because examples of the opposite also exist in which the young are more resistant than the old. By using aged or senescent animals examples of chemical toxicity have been identified in which the old exhibited either a greater sensitivity or a greater resistance than younger adults. The unpredictable nature of such differences is what makes toxicity testing not only desirable but essential.

The general state of health of the animal, pre-existing diseases, or damage to a particular organ can greatly influence the LD50 determination. Liver damage or disease can lead to an increase or a decrease in toxicity (see Chapter 7). Poor kidney function may slow the body's ability to excrete the chemical leading to its accumulation and the expression of toxicity.

Differences in toxicity are sometimes associated with differences in the diets fed to the animals or to the adequacy of the diet fed. When rats were fed diets containing no protein (in the form of casein) for 28 days after weaning, the acute oral LD50 of the fungicide, captan, was 6 ± 3 mg/kg, but when the diet contained 26% casein, the acute oral LD50 was 12,000 ± 2100 mg/kg.

The route by which the animal is exposed to the chemical can be particularly important. The chemical can be given by mouth, applied to the skin, injected by various routes, or inhaled via the lungs. In Chapter 1 it was pointed out that curare was dangerous when introduced into the blood stream, but relatively nontoxic when taken by mouth. Administration by different routes may yield different LD50 values, but in some cases the qualitative nature of the toxic response may change dramatically. In general, chemicals are more toxic when given by routes that most rapidly increase the blood concentrations, e.g., intravenous injection or inhalation of a gas. An interesting exception is the naturally occurring glycoside (a chemical containing a sugar attached to a nonsugar part), amygdalin, which is the major ingredient in the quack cancer remedy, laetrile. Amygdalin is found in peach and apricot pits, and it can be broken down by plant enzymes also found in those seeds to cyanide. Such enzymes do not exist in animal tissues, but they are found in the bacteria that normally live in the intestinal tract. For this reason, amygdalin is an unusual example of a chemical that can be as much as 40 times more toxic when given by mouth than when given by intravenous injection.

The ambient temperature can also be a factor in determining sensitivity to toxic effects. Some chemicals are more toxic on hot days. Dinitrophenol was once used in munitions, and it is still occasionally encountered as an herbicide, insecticide, fungicide, and wood preservative. Occupationally exposed workers in temperate climates showed more signs of toxicity in the summer than in the winter. In contrast, alcohol, by facilitating body heat loss, is far more dangerous when one is exposed to the cold. Housing conditions, seasonal variations, time of day or night (diurnal rhythms), and a host of other factors have also been shown in specific cases to influence the LD50. Appropriate controls can be instituted in the experimental design for any of the above sources of variation in any given laboratory. Problems arise immediately however, in attempting to institute controls between or among laboratories.

A great deal of important ancillary information can be gathered at the same time that LD50 testing is done, which adds enormously to its numeric value. For example, the time interval between dosing and death is an extremely important observation. This so-called latent period may vary widely from chemical to chemical. Death caused by rapidly acting cyanide can occur within a few minutes if it is inhaled as a gas, and it is rarely delayed for more than an hour if it is taken by mouth. Recovery from nonfatal cyanide poisoning is also rapid. Liver damage after a large overdose of acetaminophen (i.e., Tylenol) may not become evident for 1 or 2 days. Damage to the bone marrow after exposure to benzene may not be recognized for a week or two. Cancer, when caused by chemicals, may be thought of as a special kind of toxic effect and it may take months or years to develop after the exposure.

Therefore, it is important to report the time span over which the animals were observed after dosing, e.g., a 24-hour LD50 or a 7-day LD50. It is a matter of common sense that the period of observation reported should be the shortest interval in which most or all of the deaths (or other endpoints) occurred, even though the animals may be observed for longer periods of time to guard against the possibility of late events. It is misleading to report a 7-day LD50 if all of the animals died within the first hour, but it could be even worse to terminate an experiment at 24 hours when many deaths might have occurred on the third day.

The trained observer would also be taking note of the signs of intoxication in the poisoned animals. Signs are what the observer sees; symptoms are what another person tells you he or she is experiencing. Does the chemical produce tremors and convulsions indicating that it is a brain stimulant, or does it result in sleep or coma indicating that it is a depressant? Sometimes it is easy to tell whether the heart stops beating before breathing movements stop or vice versa. Such an observation may indicate that the chemical primarily affected the heart or had some primary effect that arrested breathing. The animal would be examined internally both by ordinary visual inspection (gross autopsy) and by microscopic examination of slides prepared from thin tissue slices (histologic examination). The objective would be to try to establish the cause of death by determining the poison's target organ or organs. Was there evidence of liver damage, kidney damage, or damage to any organs or tissues? Although the LD50 alone has many limitations, a wealth of other important toxicologic information can be gathered at the same time.

It may be of interest to look at the range of toxicities of tested chemicals because that is one use to which LD50 data can be put. One arbitrary system for grouping chemicals on the basis of their acute toxicity is shown in Table 3–1. Chemicals with a toxicity rating of 1, "practically nontoxic," would include such materials as talc, graphite (as in "lead" pencils), and clays such as bentonite and kaolin. Although the table indicates that the acute oral LD50 for this group in adult humans is greater than 15 gm/kg, in point of fact these materials are not absorbed from the gastrointestinal tract, and, therefore, they are incapable of producing systemic poisoning. Their administration in huge amounts may be hazardous in that it might result in intestinal obstruction or rupture, but they are not toxic.

TABLE 3–1. A TOXICITY RATING CHART FOR CHEMICAL INGREDIENTS IN AMERICAN CONSUMER PRODUCTS

Toxicity Rating or Class	Probable Oral Lethal Dose (Human)	
	Dose	For 70 kg Person (150 lb)
6—Supertoxic	<5 mg/kg	A taste (less than 7 drops)
5—Extremely toxic	5–50 mg/kg	Between 7 drops and 1 teaspoonful
4—Very toxic	50–500 mg/kg	Between 1 tsp. and 1 ounce
3—Moderately toxic	0.5–5 g/kg	Between 1 ounce and 1 pint (or 1 lb)
2—Slightly toxic	5–15 g/kg	Between 1 pint and 1 quart
1—Practically nontoxic	>15 g/kg	More than 1 quart (2.2 lb)

(After Gosselin RE, Smith RP, Hodge HC: Clinical Toxicology of Commercial Products, 5th ed. Baltimore, Williams & Wilkins, 1984.)

At the other extreme, Table 3–1 defines a toxicity rating of 6, "supertoxic," as including chemicals with an acute oral LD50 in adults of less than 5 mg/kg. This schema was devised to classify the toxicities of ingredients in American consumer products. For obvious reasons few consumer products contain ingredients with a toxicity rating of 6. Chemicals that would fall into this class include cyanide and the professionally used insecticide, parathion. Dangerous as these chemicals are, they are nowhere near the extremes represented by the most toxic chemicals known to man. As already indicated the LD50 of TCDD in the guinea pig may make it the most toxic synthetic chemical yet discovered. The most toxic of all chemicals known are naturally occurring. They are manufactured and secreted by a bacteria, *Clostridium botulinum.* These are a family of protein toxins that are responsible for a rare but often fatal form of food poisoning called botulism. A dose of 0.1 nanogram (0.1×10^{-9} g) is sufficient to kill a mouse. If humans are equally sensitive, a lethal dose may be on the order of 2.3 mg. When the so called "A" toxin is injected into mice, it produces death within a few hours. These toxins are at least 100,000,000 times more toxic than the upper limit for a toxicity rating of 1 (Table 3–1). Thus, a broad range exists for the acute toxicities of chemicals of all types, but most synthetic chemicals found in American consumer products would cluster around toxicity ratings 3 to 4.

Another use for the LD50 is to test for interactions between two chemicals in the body. When two chemicals are given together or in relatively close proximity in time so that they are present at the same time in the body, they may interact in three possible ways: (1) one may antagonize the toxic effects of the other (act as an antidote), (2) the chemicals may exhibit simple additive toxicity, or (3) one chemical may increase (potentiate) the toxic effects of the other to a point above and beyond simple addition. This latter effect is sometimes referred to as synergism.

Rational specific antidotes are known for only a small fraction of all the known toxic chemicals, but each that has been discovered represents both a toxicologic milestone and at least one important principle. It is therefore appropriate that we discuss one example in detail. Figure 3–1 shows a family of six dose-response curves, all plotted on the same abscissa with the dose expressed in mmol/kg so that they can be compared directly. This is another advantage of expressing doses in these units instead of absolute weights (see Chapter 2). The open symbols show the dose-mortality relationships for three simple inorganic chemicals of relatively high toxicity, sodium cyanide, sodium sulfide, and sodium azide. They are all in the form of sodium salts, which dissolve in aqueous solutions and separate (hydrolyze) into sodium cations and the respective anions. Thus, any difference in toxicity must be caused by differences in the respective anions, cyanide (CN^-), sulfide (HS^-) and azide (N_3^-). Clearly, cyanide is the most toxic in that it has the smallest LD50 by a factor of about five, whereas sulfide and azide are about equitoxic. Another simple inorganic chemical, sodium nitrite, was already known to be an antagonist or antidote to cyanide. The question was whether or

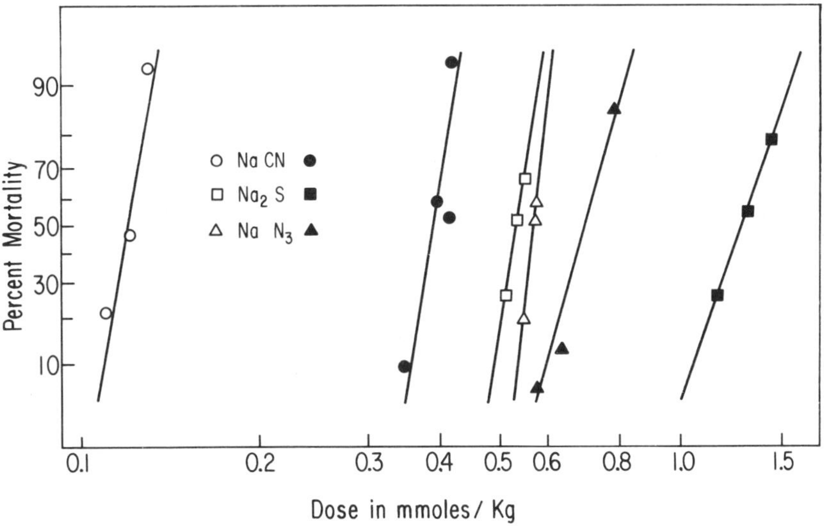

Figure 3–1. The dose-mortality relationships in mice for the sodium salts of cyanide, sulfide, and azide as given alone and as given after a constant dose of sodium nitrite at 1.1 mmol/kg. (After Smith RP, Gosselin RE: On the mechanism of sulfide inactivation by methemoglobin. Toxicol Appl Pharmacol 8:159, 1966.)

not nitrite might also be an antagonist to sulfide or azide. The solid symbols of the same shape show what happened when the three poisons were given to mice shortly after sodium nitrite, the alleged antagonist, was given.

The dose-mortality curve for cyanide given after nitrite was shifted to the right in a parallel fashion. The LD50 for cyanide was increased by about 3-fold by nitrite pretreatment. Some of the animals still died from cyanide poisoning, but it now took larger doses of cyanide to produce an equivalent percentage of mortality in the presence of the antagonist. Thus, nitrite has a significant, but limited, antidotal capacity against cyanide. Such is also the case with all antidotes known. If the dose of the antagonist is increased, the antidotal effect may increase, but there is always a limit imposed by the toxicity of the antagonist, which is also a toxic chemical. In accordance with the Paracelsus Principle (Chapter 1), everything is a poison, and only the dose determines whether the effect will be beneficial or harmful. The dose of nitrite shown in Figure 3–1 as an effective antidote to cyanide is 1.1 mmol/kg, but the LD50 for nitrite itself given alone under the same circumstances is 2.3 mmol/kg. The slightly greater than two-fold factor (potency ratio) in dose changes the chemical from an antidote to a lethal poison.

The dose-mortality curve for sulfide was also shifted to the right by nitrite pretreatment as was the curve for azide. The LD50 for sulfide was increased by something over two-fold, but that for azide was increased only by a little over one-fold. Thus, nitrite has antidotal effects against sulfide and azide as well, but they were not antagonized to as great an extent as was cyanide. The antidotal capacity can be formally expressed as a potency ratio (PR):

$$PR = (LD50\ CN^{-}\ \text{with nitrite})/(LD50\ CN^{-}\ \text{without nitrite})$$

The actual PR was 3.0 for CN^{-}, 2.2 for HS^{-}, and 1.2 for N_3^{-}. Although it is not obvious from simple visual inspection of Figure 3–1, all six curves can be proved mathematically to be parallel within experimental error. Indeed, that is a necessary condition for an analysis by PR, and it is a condition that is not always obtained experimentally. This experiment was designed to show the limits to the antidotal capacity of nitrite against each of the three toxic anions above, but it should be obvious that in each case there will be a range of doses for each of the anions over which nitrite pretreatment will allow all of the animals to survive, i.e., the antidote will be 100% effective. That range of doses will be widest in the case of cyanide and narrowest in the case of azide. At the same time, doses that lie beyond the highest dose given for each of the anions after nitrite are likely to produce 100% mortality, i.e., the antidote will no longer have any effect.

The fact that all three anions have parallel dose-mortality curves and a common antagonist that results in parallel shifts in each of those curves is consistent with the hypothesis that they all have the same toxic mechanism of action, but it by no means proves the point. The only absolute proof of the hypothesis would be to establish unequivocally the toxic mechanism of action for cyanide, and then prove that sulfide and azide do the same thing. In fact, it is now known with reasonable certainty that sulfide and cyanide have the same toxic mechanism of action, but that azide appears to act by a very different mechanism.

Another antagonist to cyanide is also recognized, namely sodium thiosulfate, which has an entirely different mechanism of action from that of nitrite. Potency ratios (PR) can be used to evaluate how effective these two antagonists are if they are given together and can be compared with their effects when given separately. Under somewhat different conditions the PR for nitrite alone against cyanide was 2.6 whereas the PR for thiosulfate alone was 4.0. When the two antagonists were given together the actual experimental PR was 6.1, which is close to the arithmetic sum of the PR for the two antidotes given alone, namely 2.6 + 4.0 or 6.6. Thus, the two antagonists can be said to have additive antidotal effects within experimental error. Had it been the case instead that nitrite potentiated the toxicity of cyanide (synergism), that dose-response curve would have been shifted to the left instead of the right. Because cyanide and sulfide are known to have the same mechanism of action, they would probably exhibit additive toxicity. Therefore, a combination of one-half LD50 of cyanide and one-half LD50 of sulfide as given together would probably be lethal to 50% of a population of animals.

One an also test for these kinds of interactions by using a special graph called an isobologram. One example is shown in Figure 3–2. A test was conducted for possible interactions between two chemicals, A and B. One way of proceeding is to determine first the LD50 for each chemical as given alone. These values are plotted on the ordinate and on the abscissa; it is immaterial which one goes where. The scales on the two axes are also immaterial. In Figure 3–2 one scale has units ten times larger than the other, but any convenient scale will do. The two values are then connected by a straight line which represents the line (I) of strict additivity of toxicities as in the example of sulfide and cyanide above. Next, the LD50 of chemical A was determined as given together with a constant dose of one-half LD50 of chemical B. The LD50 of chemical B was next determined as given together in all cases with the constant dose of one-half LD50 of A, and these two points were plotted on the isobologram. If these two points had fallen below the line of additivity (I) in region II, it would indicate a synergistic effect between the two chemicals, or a greater than

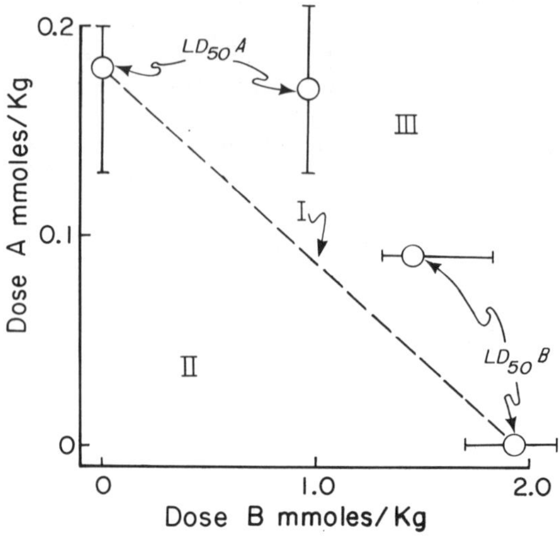

Figure 3–2. An isobologram illustrating antagonism between chemical A and chemical B. (See text for details.) (After Felsenstein WC, Smith RP, Gosselin RE: Toxicologic studies on the avicide, 3-chloro-p-toluidine. Toxicol Appl Pharmacol 28:110, 1974.)

additive toxicity. It would not be possible to tell from that observation alone if one chemical was potentiating the toxic effects of the other, or if they had mutually potentiating effects. Instead, however, as shown in Figure 3–2, the points fell outside the triangle of potentiation (II) and above the line of additive toxicity (I), in region III. This result indicates an antagonistic interaction between the two chemicals although again it is not possible from that result alone to tell if one chemical is antagonizing the effects of the other or if they have a mutual interaction.

Another example of an isobologram, this one illustrating potentiation, is shown in Figure 3–3. Here two drugs that are each useful alone in treating malaria have been tested together in various proportions. The ED50 is defined here as the dose that reduced the number of malarial parasites in the blood of infected chicks to 50% of the number in untreated control chicks. The two drugs act by different mechanisms, and it can be seen that the various combinations are effective at much lower doses of the two ingredients than the doses required for each as given alone.

From the above example it can be seen that if only two ingredients are present together in a mixture and if each can be obtained alone in pure form, methods exist for evaluating the acute toxicity of the mixture and for testing for interactions between the two ingredients. For mixtures containing more than two chemi-

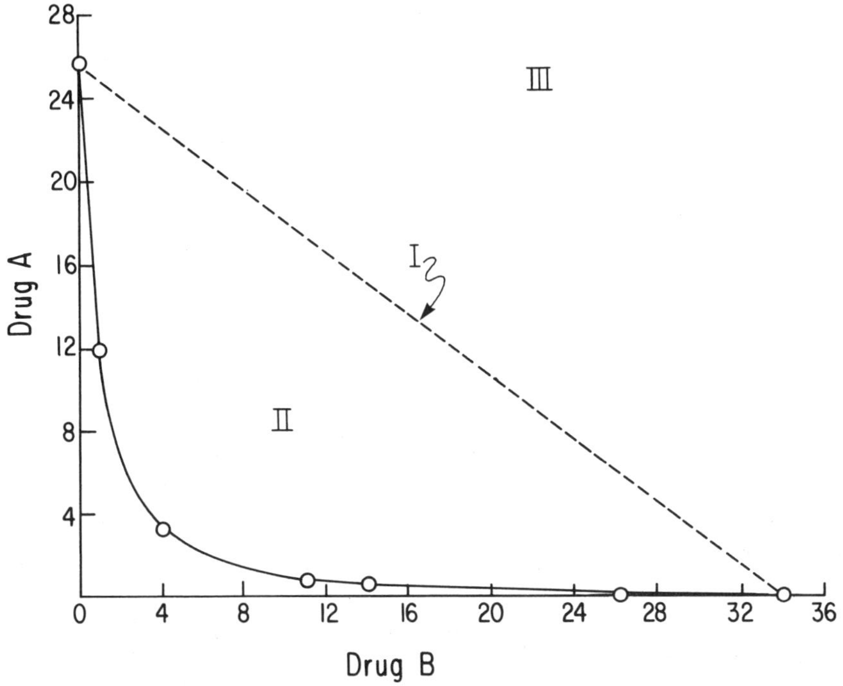

Figure 3–3. An isobologram illustrating synergism (potentiation) between chemical A and chemical B, each of which is useful alone in the treatment of malaria. (See text for details.) (After Rollo IM: The mode of action of sulfonamides, proguanil, and pyrimethamine on Plasmodium gallinaceum. Br J Pharmacol 10:208, 1955.)

cals there are two less satisfactory approaches for evaluating the toxicity of the combination. In mixtures in which all the ingredients are known with certainty, one approach is to assume that the toxicity of the mixture is caused solely by its most toxic ingredient. All other ingredients are regarded as inert. In adopting this approach another relevant fact is needed, namely, the concentration of the most toxic ingredient in the mixture. If a mixture contains a very toxic chemical but it is present in only vanishingly small amounts, that is different from when the most toxic ingredient makes up the bulk of the mixture. One purely empirical way of making those sorts of adjustments can be illustrated in the case of a mixture containing cyanide as the most toxic ingredient. As already noted cyanide has a toxicity rating of 6. A rule of thumb is that if the mixture contains 10% or more cyanide, the toxicity rating of the mixture is also 6. If the mixture contains between 1 and 10% cyanide, however, the toxicity rating drops to 5, and if it contains between 0.1 and 1% cyanide the rating becomes 4. This approach is obviously only the crudest sort of a guess, which totally ignores any possibility for complex interactions among the ingredients of the mixture.

The alternative to the above is to regard the mixture as simply a single entity, and to determine its LD50 directly. Although perhaps more accurate, this approach is not applicable to mixtures in general, and as encountered in the environment, no two mixtures are ever alike. Thus, the LD50 for one mixture has little or no relevance to another. Moreover, even the composition of any one mixture in the environment is apt to vary with time as the more volatile ingredients are lost to the atmosphere, as ingredients react chemically with each other, as ingredients are separated from each other during seepage through soil, and as a host of other factors occurs.

Returning to the subject of potency ratios (PR) and the experimental results of Figure 3–1, it was found that the PR for the combination of cyanide and nitrite was 3.0 relative to cyanide alone, whereas the PR for azide and nitrite was 1.2. Because the latter PR is so small, it is not clear whether the effect on azide was truly caused by nitrite, i.e., was it "statistically significant," or whether it was merely a result of chance experimental error? The usefulness of the PR (as well as the ED50 or LD50) is improved greatly if its 95% confidence limits (CL) are also computed. If the 95% CL around a PR does not include the value of 1.0, the treatment is said to be significantly different from the control group at the 5% level of confidence ($p < 0.05$).

How PR and its CL can be put to practical use is illustrated in Table 3–2. It has already been noted that nitrite is an antagonist to both cyanide and sulfide, and that thiosulfate is also an antagonist to cyanide. Oxygen in combination with one or both antagonists is said to add to the antagonistic effects although it has

TABLE 3–2. EFFECT OF VARIOUS PRETREATMENTS ON ACUTE SULFIDE POISONING

Pretreatment	LD50 (mmol/kg)	95% CL	PR	95% CL
Air	0.25	0.23–0.27		
Oxygen	0.27	0.25–0.29	1.1	1.0–1.2
Thiosulfate	0.28	0.27–0.30	1.1	1.1–1.2
Oxygen + thiosulfate	0.30	0.28–0.33	1.2	1.1–1.4
Nitrite	0.63	0.58–0.71	2.5	2.3–2.8
Oxygen + nitrite	0.58	0.54–0.58	2.3	2.1–2.5
Thiosulfate + nitrite	0.50	0.46–0.54	2.0	1.8–2.2
Oxygen + thiosulfate + nitrite	0.58	0.58–0.63	2.4	2.2–2.6

(After Smith RP, Kruszyna R, Kruszyna H: Management of acute sulfide poisoning, effects of oxygen, thiosulfate and nitrite. Arch Environ Health 31:166, 1976.)

no effect alone against cyanide. The question is whether or not these patterns for cyanide antagonists extend also to sulfide. In testing all possible treatments alone and in all possible combinations, a series of eight dose-mortality curves was generated, and the PR with 95% CL was generated for each regimen using sulfide in air as the control treatment.

The PR in Table 3–2 for O_2 vs. air was 1.1, but the CL included 1.0; thus, the hypothesis that oxygen under these conditions is an antagonist to sulfide was rejected. The next two treatments include thiosulfate. Although the PR are the same or only slightly larger than in the case of O_2, we must accept the CL as telling us that these are small, but statistically significant effects at the 5% level of confidence. At the same time common sense tells us that those effects are so small, they are scarcely worth further investigation into their biological mechanism, i.e., they are almost "biologically insignificant." In contrast, the next four treatments all include nitrite, the PRs are double those of the previous treatments and all are statistically significant. The conclusion is that nitrite is a useful antagonist to sulfide, and that little if anything is added under these conditions by also giving thiosulfate or O_2 or both. At the same time thiosulfate and oxygen appear to do no harm because if they did, their PRs would be less than 1.0.

SUMMARY POINTS

The adverse effects expressed after a single dose or brief exposure to a chemical are referred to as acute toxicity.

Acute toxicity varies with species, sex, age, general state of health, diet, and a host of other factors.

The interval between exposure and the appearance of acute toxicity (latent period) may vary from minutes to years.

A few chemicals are virtually nontoxic because they are not absorbed, but LD50 values can vary over a 100 million-fold range.

When two chemicals are present together in the body they may interact to antagonize (antidote), potentiate, or add to each others' effects.

Antidotal capacities may be expressed quantitatively in terms of potency ratios and their confidence limits.

BIBLIOGRAPHY

Apfel RJ, Fisher SM: To Do No Harm: DES and the Dilemmas of Modern Medicine. New Haven, CT, Yale University Press, 1984.

Felsenstein WC, Smith RP, Gosselin RE: Toxicologic studies on the avicide, 3-chloro-p-toluidine. Toxicol Appl Pharmacol 28:110, 1974.

Gosselin RE, Smith RP, Hodge HC: Clinical Toxicology of Commercial Products, 5th ed. Baltimore, Williams & Wilkins, 1984.

Rollo IM: The mode of action of sulfonamides, proguanil, and pyrimethamine on Plasmodium gallinaceum. Br J Pharmacol 10:208, 1955.

Smith RP, Gosselin RE: On the mechanism of sulfide inactivation by methemoglobin. Toxicol Appl Pharmacol 8:159, 1966.

Sun M: Lots of talk about LD50. Science 222:1106, 1983.

Way JL, Gibbon JL, Sheehy M: Cyanide intoxication: protection with oxygen. Science 152:210, 1966.

Zbinden G, Flury-Roversi M: Significance of the LD50 test for the toxicological evaluation of chemical substances. Arch Toxicol 447:77, 1981.

4

CHAPTER

INHALATION TOXICITY AND THE CONCEPT OF CHRONICITY

As indicated in Chapters 2 and 3, the LD50 is a useful parameter for expressing the acute toxicity of solid and liquid chemicals for comparative purposes and for testing for interactions between chemicals. Solutions of such chemicals can be prepared, and portions of these may be given as single, discrete doses by any of a number of routes of administration to laboratory animals. The situation is different if one is presented with a problem involving the toxicity of chemicals that are inhaled via the lungs such as gases, vapors, dusts, and aerosols (stable suspensions of solid or liquid chemicals in air). In previous chapters sodium sulfide has been used as an example of a toxic chemical, but human poisonings are invariably a result of exposure to hydrogen sulfide gas (H_2S) instead of the sodium salt. The gas is encountered in sewers, cesspools, and modern facilities for processing liquid manure for use as a fertilizer. Thus, it is generated during the putrefaction of organic matter. Sulfide salts, which may give rise to H_2S gas, are used to remove the hair from hides in tanneries, and the gas is found naturally associated with deposits of "sour" gas, petroleum, and volcanic gases. It does not necessarily follow that the toxic effects of inhaled H_2S are the same as those of injected sodium sulfide. For example, the effects of inhaled cadmium fumes are exerted initially in the lungs, the effects of cadmium salts ingested by mouth are exerted primarily in the gastrointestinal tract, and the effects of injected cadmium salts are exerted primarily in the kidneys.

The field of inhalation toxicology presents special problems. The first of these relates to the time of exposure. Except for a few highly toxic gases in appropriate concentrations such as H_2S or hydrogen cyanide (HCN), which are capable of producing death within a few minutes, many gases, especially in low concentrations, require appreciable exposure times to produce toxic effects or death. The dose simply cannot be given all at once in a single instant of time as was the case in acute toxicity testing of solid or liquid chemicals.

A gas must be inhaled from the ambient environment, mix with the air in the lungs, penetrate into the deep lung structures, be absorbed into the blood, be distributed to its target organ, achieve a critical concentration in that organ and exert its characteristic toxic effect, which may or may not involve a long latent period. All of these processes require time and the latent period may be the major determinant for the appearance of toxic signs. Depending on the length of the latent period, toxic effects may occur during relatively brief periods of exposure, e.g., minutes to hours, or they may be delayed for variable periods of time following termination of the exposure, as was the case in acute toxicity testing.

One way of evaluating the toxicity of a gas is to expose animals to it continuously until all are dead (Fig. 4–1). One of the problems with this approach is that, in general, the higher the ambient gas concentration (up to certain limits) the more rapidly death is produced. For most gases there is also a safe range of ambient concentrations that can be tolerated indefinitely, a no observable effect level (NOEL). Therefore, the toxicity of a gas is both a function of its ambient concentration and of the time of exposure. For many gases, however, that relationship cannot be expressed by any simple mathematical formula.

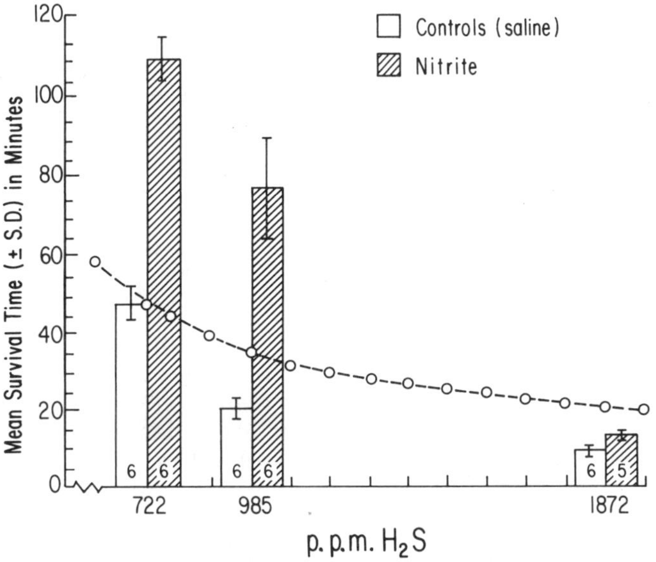

Figure 4–1. Mean survival times of mice exposed to various concentrations of hydrogen sulfide in air. The vertical lines indicate ±1 standard deviation. Open bars show results obtained with groups of control mice. Hatched bars show results obtained with mice that had been pretreated with sodium nitrite prior to exposure. The dashed curve indicates a rectangular hyperbola computed using the results obtained at 722 ppm in control mice as a reference point. (See text for additional details.) (After Smith RP, Gosselin RE: The influence of methemoglobinemia on the lethality of some toxic anions. II. Sulfide. Toxicol Appl Pharmacol 6:584, 1964.)

The second new problem is that one almost never knows the total dose that has been absorbed systemically. It is possible to make that determination experimentally at least in some cases, but it involves far more elaborate and expensive equipment and protocols than are in general use today. Only part of the total amount inhaled is absorbed into the blood. Of that fraction which is absorbed, a variable proportion may or may not be exhaled and lost from the body. Clearly, the dose of a gas cannot be expressed simply in terms of mg/kg body weight as in acute toxicity testing. In practice it is common to express exposure in terms of the ambient concentration of gas, e.g., milligrams of gas or particulates per cubic meter of air (mg/m^3), volumes of gas per 100 volumes of air (vol%), or parts per million (ppm) for a specified period of time.

A ppm of a gas means one part of gas by volume in one million parts of air by volume at 25°C and at a pressure of 760 mmHg. Because one gram molecular weight (mol) of an "ideal" gas occupies 22.45 liters at 25°C and 760 mmHg (equivalent to 22.4 liters at 0°C or 273°K and 760 mmHg, the so-called standard conditions of temperature and pressure), it follows that:

$$1 \text{ ppm} = (\text{mg/l}) \times (22.45 \text{ l/mol. wt.})$$

Thus, for 1.4 mg/l of H_2S:

$$\text{ppm} = (1.4 \text{ mg/l}) \text{ x } (24.45/34.08 = 1.004$$

Because there are 1000 l in 1 m^3, it also follows that:

$$\text{mg/m}^3 = (1.4 \text{ mg/l}) \times (1{,}000 \text{ l/m}^3) = 1{,}400 \text{ mg/m}^3$$

Confusion may arise because the term, ppm, is used also to indicate the concentration of a solid or liquid chemical when dissolved in water. This is common in testing chemicals for toxicity to aquatic lifeforms, such as fish, or insect larvae. In that case 1 ppm means 1 mg of chemical as dissolved in 1 l of water or any equivalent multiples or fractions of those units. In this type of testing, the organisms are immersed continuously in a solution of the chemical so that again precise specification of the duration of the exposure is critical. Otherwise, the test system is in most ways analogous to gas exposures. Finally, the same term, ppm, may be used in feeding experiments (Chapter 5) in which the chemical is mixed uniformly in the feed. In this case 1 ppm would be equivalent to 1 mg of chemical per kg of feed.

Intuitively, it might seem as if the product of a gas concentration and the time of exposure to a given endpoint should be a constant:

$$C \times T = K$$

which is recognizable as the formula for a rectangular hyperbola. Many gases conform approximately to this model, but at least an equal number do not. Those that conform tend to be irritant gases that affect primarily the lungs, e.g., formaldehyde, nitrogen oxides (NO_x), sulfur dioxide (SO_2), or chlorine (Cl_2). Gases that produce systemic poisoning such as HCN or H_2S (Fig. 4–1) tend not to fit the model. In order for the model to hold, the gas would have to exhibit almost strictly cumulative toxic effects at all concentrations. Such is almost never the case with gases that produce systemic poisoning because a variable fraction of the amount absorbed is inactivated by the body before it can exert its toxic effect. Moreover, most gases have a definable NOEL, which could not be true if they had absolutely cumulative effects. At the NOEL, inactivation is presumably keeping pace with the rate of absorption so that the gas does not accumulate, and toxic signs never appear. Because the rectangular hyperbola model is not generally applicable, it is now obsolete.

Whatever experimental protocol is used, it is vital that the time of actual exposure be explicitly specified in any final report. Emphasis must be placed on ensuring that the animals are exposed to pure uniform ambient concentrations of the gas. The design and operation of satisfactory exposure chambers is an engineering technology unto itself.

Figure 4–1 summarizes the results of an experiment in which mice were exposed continuously to toxic concentrations of H_2S until all the animals in each group were dead. The survival time was noted for each mouse, and at the end of the experiment, the average or mean survival time for each group was calculated together with its respective standard deviation. The open bars in Figure 4–1 indicate the results obtained with control or untreated mice. As expected, the mean survival times became shorter as the gas concentration increased. The mean survival times, however, were not a good fit for the rectangular hyperbola model above (dashed curve in Figure 4–1).

Other groups of mice were pretreated with sodium nitrite, as described in Chapter 3, and exposed to the same gas concentrations (hatched bars in Figure 4–1). In each case the pretreated mice survived significantly longer than their comparable control group. Thus, nitrite is an antagonist to inhaled H_2S just as it was an antagonist to injected Na_2S. This observation suggests, but does not prove, that the effects of the inhaled gas are identical to those of the injected salt. As already noted, different routes of administration of the same chemical may give rise to different forms of poisoning, and the same route of administration of different forms of the same chemical may do so as well. In this experiment we have given different forms of the same chemical by different routes with essentially the same result.

The survival time also decreased as the gas concentration increased in nitrite-pretreated animals (Figure 4–1). This result reinforces the concept that nitrite has a significant but limited capacity to antagonize sulfide whether it is given as a single dose or as a continuous gas exposure over many minutes. It is also of interest that the greatest difference in survival time between the control animals and the nitrite-treated animals occurred at the intermediate concentration of H_2S tested, 985 ppm. The difference in survival time was small at the highest gas concentration because the limited antagonistic capacity was quickly overwhelmed (it quickly reached saturation). Had the H_2S concentration been increased further, the difference between control and treated groups would likely disappear altogether. If, on the other hand, the gas concentration were decreased below 722 ppm, at some point a concentration would be reached at which even the control mice could tolerate the exposure indefinitely (NOEL), nitrite would be superfluous as an antagonist, and again the difference between control and treated groups would disappear. The determination of such safe levels of exposure that can be tolerated indefinitely by normal individuals is the goal of industrial hygienists striving to protect the worker in occupational settings.

In an alternative experimental design the time of exposure is held constant, and the gas concentrations are adjusted such that some but not all of the animals respond at each concentration. If at least three gas concentrations are tested one could derive from the results a parameter analogous to an LD50, namely, a calculated gas concentration that would be lethal to 50% of a population of animals during (or after) the specified exposure time. This parameter could be reported, for example, as a 7-day LC50 for a 1 hour exposure. The distinction between "dose" and "concentration" is a fundamental one in practical terms, but biologic responses plotted against the log of either yield basically the same kind of relationship. The parameter, LC50, is subject to all the same limitations and ambiguities as the LD50 together with some new complications. For example, if the whole animal is exposed to the gas, one can never be entirely certain that the route of absorption was limited exclusively to the lungs. Some material may be absorbed through the skin, and some may be licked off the fur. The rate of breathing of the animal may change during the exposure so that the rate of absorption is altered. Some gases stimulate the rate of breathing, other decrease it, and still others do both over the course of the exposure. These are the kinds of variables and considerations that always confront the investigator who is testing for toxicity in whole animals.

As already noted, there are certain similarities between inhalation toxicology testing and toxicity testing using aquatic species. In the latter, the organism is continuously immersed for a specified exposure period in an aqueous environment in which the solid or liquid chemical is dissolved. The larvae of mosquitos that transmit yellow fever have been used for some years to test for the toxic effects of insecticides intended for use against them. It was recently suggested that the same system might be more generally applicable to predict the acute toxicities of chemicals for mammalian species. Larvae in distilled water were exposed to various concentrations of chemicals for defined periods of time, e.g., 0.5, 1 or 24 hours, and LC50 values for a constant exposure time were calculated using probit analysis. It is likely in these experiments that the route of exposure was some undefined combination of ingestion and direct penetration through the outer covering of the larvae.

The LC50 values (in ppm) derived in this way were compared with acute oral LD50 values in rats (mg/kg) for a small series of chemicals. As might be expected for two different routes of exposure in two different species, the absolute values differed widely, sometimes by as much as 2 to 3 orders of magnitude (100- to 1000-fold). However, a significant correlation existed between the mosquito larvae LC50 and the rat LD50 values. Thus, if only the rank order of toxicity is under consideration, the mosquito larvae bioassay has significant advantages in terms of economy.

Workers in many industries are exposed to a variety of airborne gases, vapors, dusts, and fumes. Efforts have been ongoing for many years to establish standards or limits for ambient concentrations of these materials in the workplace. The goal has been to try to define concentrations that would be safe even with repeated daily exposures over the entire productive lifespan of the worker. An organization known as the American Conference of Governmental Industrial Hygienists (ACGIH) has set such limits for about 600 chemicals, and periodically reviews them as new data accumulates. These are known as threshold limit values or TLV.

The TLV are set for airborne concentrations of chemicals to which workers, of any age, sex, or body weight may be safely exposed for 8 hours per day, 5 days per week, over their occupational lifetime. The basis for setting a TLV is not always systemic toxicity. The values may be set on the basis of eye irritation, as in the case of H_2S, or on the basis of skin irritation, odor nuisance, or other criteria. A periodically revised volume, *Documentation of the Threshold Limit Values,* indicates the basis on which each TLV was set.

In many industrial settings the ambient concentrations of any given chemical may vary widely during the day depending on such factors as the rhythm of the particular operation, differing conditions of ventilation, and changes in production schedules.

A practical way of dealing quantitatively with these fluctuating concentrations is to express the TLV in terms of a time-weighted average (TWA) concentration over an 8-hour period. These averages may be calculated from periodic (instantaneous "grab") samples of the workroom air or from the results of continuous monitoring. In either case it is inevitable that some samples or some time periods will show occasional excursions above the TLV, whereas other samples or time periods will show that the TLV is being met.

Provisions for unusually large excursions above the TLV have been addressed by the ACGIH, and some rules of thumb have been formulated. For example, for the group of chemicals having a TWA-TLV between 100 and 1000 ppm any instantaneous excursion should not exceed the TWA-TLV by a factor of more than 1.25. Such provisions are aimed at eliminating excessively high concentrations over short intervals of time, which when expressed as a TWA might show conformance to the TLV.

Some TLV are set only as ceiling values (CV) because the harmful effects of even short term excursions above the CV are so undesirable or dangerous that they outweigh any possible long-term or chronic effects. The protection of workers exposed to such chemicals depends on the judgement of the ACGIH in keeping the ambient concentrations below the CV at all times. An example of such a chemical is an acid mist because even short excursions above the CV could result in serious eye injury.

When TLV are set on the basis of systemic toxicity, available data on long-term, low-level exposures are often scant and usually of three different kinds: human experience in the work place or data from human volunteers, animal testing, or analogy with similar chemicals. As noted above, the setting of TLV is an ongoing process, and various TLV may be revised upward or downward as more reliable information accumulates. Table 4–1 lists some TLV, CV, and short-term exposure limits (STEL) (see below) for commonly encountered industrial chemicals. The TLV vary widely from the more hazardous chemicals (ozone, formaldehyde, hydrogen cyanide, hydrogen sulfide, and sulfur dioxide) to the less hazardous ones (carbon dioxide, ethyl alcohol, and liquid petroleum gas). Liquid petroleum gas sounds like a contradiction in terms. As marketed it is compressed into a liquid form in tanks, but the TLV is set for the gaseous form. The material is essentially devoid of systemic toxicity, but like any nontoxic, inert gas it is capable of producing asphyxia if it is present in concentrations so high that it displaces oxygen below critical levels. Its major hazard is its flammability. The TLV for carbon tetrachloride and phenol were set on the basis of hazard to the skin.

TABLE 4–1. THRESHOLD, CEILING, AND SHORT-TERM EXPOSURE LIMIT VALUES FOR SOME COMMONLY ENCOUNTERED ENVIRONMENTAL CHEMICALS*

Substance	ppm	mg/m³
Carbon dioxide	5000	2000
Carbon monoxide	50	55
Carbon tetrachloride (skin)	5	30
Cresol	5	22
Ethyl alcohol	1000	1900
Formaldehyde	1	1.5
Hydrogen cyanide (ceiling limit)	10	11
Hydrogen sulfide	10	14
Hydrogen sulfide (short term exposure limit)	15	21
Liquified petroleum gas (LPG)	1000	1800
Ozone	0.1	0.2
Phenol (skin)	5	19
Sulfur dioxide	2	5

*Expressed as parts of vapor or gas per million parts of contaminated air by volume at 25°C and 760 mmHg pressure or as approximate mg of particulate per cubic meter of air. (After The American Conference of Governmental Industrial Hygienists Inc.: Documentation of the Threshold Limit Values, 4th ed. ACGIH, Cincinnati, OH, 1980.)

In some industrial operations exposures occur in a predictable manner, but they are of brief duration, relative to an 8 hour shift. In such cases a TWA-TLV may not be relevant, and short term exposure limits (STEL) may be set for time periods of 5, 15, or 30 minutes. These are again TWA but for shorter exposure periods with the provision that the average 8-hour exposure under the STEL will not exceed the TWA for an 8-hour exposure under the TLV.

A great deal of unvarnished guess work goes into setting CV, STEL, and especially TLV because data relevant to lifetime, low-level exposures and including actual, reliable measurements of the ambient concentrations are scant or nonexistent for many chemicals. The carcinogenic hazard of asbestos was not recognized for many years, and the neuropsychiatric hazards of carbon disulfide in the viscose rayon industry were recognized only after workers had been overexposed.

Inhalation toxicology represents a grey area because no generally accepted definition of what constitutes an acute exposure comparable to single-dose, acute toxicity exists. On the other hand, most would agree that chronic testing should be carried out over the entire lifespan of the species being used. In terms of environmental pollution, chronic toxicity testing is certain to yield more relevant results. Exposure to lead, for example, in any industrialized society is more or less continuous over the human lifespan. In designing experiments to evaluate chronic toxicity, one must think in terms of 70 years for humans, 10 years for dogs, and 2 years for rats. Even with the relatively short-lived rat, such testing is tedious, time-consuming, and expensive. Only one chemical can be tested in any one group of animals, and such tests cost hundreds of thousands of dollars (see Chapter 5).

If the objective is to evaluate true chronic toxicity, you must actually do the experiment. Because of the time and expense, however, toxicologists have struggled for years to devise satisfactory tests involving shorter time periods in the hope that these would be predictive for the chronic case. An empirical suggestion that has been widely adopted in one form or another is based on the premise that the important toxic effects of most chemicals would be obvious within one-tenth of the lifespan if the exposure was sufficiently intense and if the tests were conducted using young growing animals. For rats the time span for testing would be on the order of 90 days. Such 90-day studies are referred to in the jargon as "subacute," although semantically "subchronic" might be more appropriate.

One special variant of subchronic testing is not required by any government agency and, therefore, it is seldom performed. However, the concepts on which it is based are highly instructive and well worth some study. In this test animals are given the same single dose of the chemical on each of 90 consecutive days. When such an experiment is correctly done, one can extract from the end results data appropriate for calculating a 90-dose LD50 in mg/kg per day, which can then be compared with the single dose LD50 in mg/kg. The ratio between these two values reveals something about the inherent capacity of the chemical to produce chronic or cumulative effects. The ratio is called the chronicity index (CI):

CI = (1-dose LD50, mg/kg)/(90-dose LD50, mg/kg per day)

From the above it should be intuitively obvious that the larger the CI, the greater the tendency for a given chemical to produce chronic toxicity. For a hypothetical, perfectly cumulative poison, the CI would be 90. If one took the single dose LD50 for such a chemical and divided it into 90 equal parts and gave one of those parts on each of 90 consecutive days, the end result would be equivalent to the 90-dose LD50. For such a hypothetical chemical the implications are that virtually every molecule would be absorbed and exert an irreversible toxic effect. No such chemical is known to exist, but a few have extreme degrees of chronic toxicity. For a chemical that has little or no tendency to produce chronic toxicity, the CI would be in the range of 1 to 2. The implication here is that the animals will tolerate a daily dose just under the single-dose LD50 for a period of 90 days or longer.

Actual CI data have been generated for only a small fraction of the chemicals known to man, but the four examples in Table 4–2 are sufficient to illustrate the important principles. As already indicated, sodium cyanide is a chemical with a high derived acute toxicity. The 90-dose LD50 for NaCN in Table 4–2 is admittedly a hypothetical value instead of an experimentally derived value, but it is an approximation supported by a large body of experience and closely related data. As indicated by its CI, cyanide has almost no tendency to produce chronic toxicity. Indeed, in an actual feeding experiment, as opposed to a rigid 90-dose experiment, the estimated CI was less than one. Penicillin is a familiar example of a chemical with low acute toxicity and little tendency to produce chronic toxicity. Mirex is a "hard-type" insecticide distantly related to DDT (see Chapter 10). It has moderately high acute toxicity, but a very high CI, which ranks it two-thirds of the way toward becoming the perfectly cumulative poison. Warfarin is a widely used rodenticide (rat and mouse poison) and also a therapeutically useful drug to prevent the complications of excessive or inappropriate blood clotting in the body (an anticoagulant). It has a high acute toxicity and a moderate CI.

The point of Table 4–2 is that no simple relationship exists among chemicals between their acute and their chronic toxicities. Many people are reluctant to believe that a highly toxic chemical, such as cyanide when given acutely, has almost no tendency to produce chronic toxicity. One might make an analogy here between the chronic toxicity of cyanide and a person standing in a pool of water which just reaches the chin. The person might remain there for very long periods of time and experience only minor wrinkling of the skin. If, however, the water level is raised only a few inches and for only a few minutes (acute toxicity of cyanide), rapid demise results.

Some chemicals are inherently likely to produce cumulative effects. The two examples of such chemicals in Table 4–2 illustrate

TABLE 4–2. DIFFERENCES BETWEEN THE ACUTE AND CHRONIC TOXICITY OF VARIOUS CHEMICALS AND THEIR CHRONICITY INDICES

Chemical	1-Dose LD50 mg/kg	90-Dose LD50 mg/kg per day	Chronicity Index
Sodium cyanide	10	9	1.1
Penicillin	6700	4440	1.6
Mirex	365	6	60.8
Warfarin	2	0.1	20.8

(After Hayes WJ Jr: The 90-dose LD50 and a chronicity factor as measures of toxicity. Toxicol Appl Pharmacol 11:327, 1967.)

the two properties that tend to result in chronic toxicity. Mirex is extremely insoluble in water, but it is highly soluble in fats and solvents that are not miscible with water such as kerosene. Mirex accumulates in the body with repeated exposures because it dissolves in and is tenaciously stored in body fat. These fat depots slowly release it and thus, accumulation in the body with repeated doses is one property that contributes to chronic toxicity. In contrast, warfarin is relatively water soluble, and the body rids itself of warfarin much more quickly than it rids itself of mirex. The half-life ($t_{1/2}$) of warfarin in blood plasma is on the order of 40 hours, i.e., the time required for the plasma concentration to decrease to one-half of its original value. Although that is a fairly long $t_{1/2}$ in terms of commonly used drugs, it is short in comparison with that for mirex and its chemical relatives. The effects of warfarin, however, persist well beyond the duration of the stay of the bulk of the chemical in the body. So, the tendency of a chemical to produce irreversible or slowly reversible effects is another property of cumulative poisons. These are often called "hit-and-run" poisons. Carcinogens are examples par excellence of such chemicals.

There are even more pitfalls in chronic toxicity testing than in acute toxicity testing. For example, in the case of both chemicals in Table 4–2 with high CI, the acute toxic effects are the same as the chronic toxic effects, and the only differences are the doses and the time required for their expression. There are numerous examples of chemicals where such is not the case. The effects of moderate doses of ethyl alcohol consumed over a short period of time consist of the familiar behavioral alterations of inebriation. Both the signs of inebriation and the alcohol itself persist only transiently in the individual. A sufficiently large single dose, e.g., a quart of whiskey consumed over a few hours, can result in a coma with subsequent respiratory failure and death. When consumed over many years in excessive amounts, each drinking bout is accompanied by alternating periods of inebriation and sobriety, but superimposed on these may be a new syndrome resulting from irreversible and, therefore, cumulative damage to the liver. The result can be cirrhosis, eventual liver failure, and death from an entirely different cause than that of the acute insult.

Other complications may arise in chronic testing because of changes in sensitivity of the animal or the target organ during the course of the exposure. As exemplified by the heroin addict, high degrees of resistance may develop to some but not necessarily all of the effects of the drug. Moderate degrees of tolerance develop to some of the effects of ethyl alcohol as well. In the case of heroin, the most important effect to which addicts become tolerant is also the most dangerous, namely, respiratory depression. Addicts may be able to tolerate as much as 10 times the lethal dose of heroin in a nonaddicted person. Both, however, will die in respiratory failure if the dose is sufficiently large. Thus, tolerance is in some ways analogous to antagonism or an antidotal effect in that it is only protective over a finite range of doses. Tolerance does not develop to constriction of the pupils, which is another characteristic of heroin and related drugs, and the pinpoint pupils will betray the user. Conversely, tolerance develops to some of the behavioral aspects of alcoholic inebriation, but not to the dose that produces respiratory failure. The phenomenon of "cross-tolerance" is recognized between heroin and morphine, in which the tolerance produced by one drug extends also to another similar one. Alcohol produces cross-tolerance with barbiturates, but cross-tolerance does not occur between heroin and alcohol. Thus, fundamentally different mechanisms are at work in the two tolerant states.

Increased sensitivity may also be exhibited in the chronic alcoholic with terminal liver disease. Alcohol is inactivated by the liver, and cirrhosis may greatly interfere with that process. In such case the usual dose of alcohol may result in higher blood levels and a longer persistence in the body than in healthy individuals. Thus, over a lifetime an alcoholic may develop tolerance and then gradually lose it to the point that he or she becomes more sensitive to alcohol than a comparable nondrinker. In the final stage the target organ (brain) has not become more sensitive to alcohol; it is simply responding appropriately to the equivalent of a larger dose and higher blood level secondary to the liver damage.

Both acute (single-dose) and chronic (multiple-dose) toxicity are determined by the size of the dose(s) and the nature of the chemical. In chronic toxicity there are additional influences such as the duration of dosing or exposure, the interval between doses or exposures, the degree of storage of the chemical in the body, the tendency of the chemical to produce irreversible or cumulative effects and changes in sensitivity of the test species over the course of exposure.

SUMMARY POINTS

The toxicity of gases, vapors, dusts, or aerosols are expressed in terms of concentration (LC50) and time of exposure instead of dose (LD50).

The safe concentrations of airborne chemicals in the workplace are expressed as threshold limit values (TLV).

Chronic toxicity refers to long-term, low-level exposures over a period of a year or more, up to the lifespan of the species used.

SUMMARY POINTS (continued)

Subchronic (subacute) studies involve periods of exposure of 90 days; often the chemical is mixed in the diet and fed at various levels.

For any given chemical no simple relationship exists between its acute and its chronic toxicity; a chemical with high acute toxicity may have a low chronic toxicity.

The characteristics that contribute to a high chronic toxicity are accumulation in the body with repeated exposure and the ability of the substance to produce irreversible effects.

BIBLIOGRAPHY

American Conference of Governmental Industrial Hygienists Inc: Documentation of the Threshold Limit Values, 4th ed. ACGIH, Cincinnati, OH, 1980.

Hayes, WJ Jr: The 90-dose LD50 and a chronicity factor as measures of toxicity. Toxicol Appl Pharmacol 11:327, 1967.

Klaassen CD, Amdur MO, Doull J: Caserett and Doull's Toxicology: The Basic Science of Poisons, New York, Macmillan, 1986.

Richie JP Jr, Mills BJ, Lang CA: The verification of a mammalian toxicant classification using mosquito screening method. Fund Appl Toxicol 4:1029, 1984.

Smith RP, Gosselin RE: The influence of methemoglobinemia on the lethality of some toxic anions. II. Sulfide. Toxicol Appl Pharmacol 6:584, 1964.

5

CHAPTER

SUBCHRONIC AND CHRONIC TOXICITY TESTING

To this point we have discussed toxicity testing only in relation to mortality. Even the concept of chronicity (see Chapter 4) was introduced in terms of a 90-dose LD50. Because lethality is the ultimate expression of toxicity, it is certainly relevant, but it is more important to know the intensity of exposure at which toxic effects begin to occur or, conversely, what is the NOEL? In order to define these limits it would be desirable to have a test that is highly sensitive, will give positive results with all kinds of toxic chemicals irrespective of their mechanism of action, is relatively simple to conduct, and can define both toxic and safe levels of exposure.

A good general test for subtle toxic effects that has been in use for many years is based on observations of weight gain in growing animals, usually rats. These experiments are more realistic than the 90-dose LD50 because the chemical to be tested is mixed with the diet or with the drinking water. Intake of synthetic chemicals accidentally or deliberately added to food or water is one of the important routes of exposure for humans. Instead of the chemical being given as a single dose at the same time each day, the exposure is spread out over time as the animals feed, which is also more in keeping with human exposure. The fact that rats are more active and feed mostly at night is certainly different from the human case, but it is probably a minor difference in terms of the results of the test.

The animals are weighed at regular intervals, perhaps daily. They are often housed in individual cages so that their food containers also can be weighed. It is necessary to follow the amount of food consumed because that permits calculation of the actual dose of the chemical consumed daily. Rats often spill and waste a certain amount and care must be taken to account for such losses. The feeding may take place over the lifespan of the species used (chronic testing which is customary in testing for carcinogenic effects) or for periods of 90 days (subchronic testing).

Because no two animals are exactly alike, at the end of the experiment it is likely that no two rats will have consumed the same amount of feed. Therefore, no two rats will have received exactly the same dose of chemical. For a given concentration of the chemical in the diet, this problem is most easily dealt with by simple averaging. The results can then be expressed as the mean dose in mg/kg per day together with its standard deviation as an indication of the variability of the results.

A well designed experiment will include several groups of animals that have been fed different dietary levels of the chemical in addition to a control group that has been fed a diet without the chemical. The goal is to define at least one level of feeding at which there was no significant effect on weight gain (NOEL level) relative to the control group, and at least one level of feeding at which there was a statistically significant decrease in the rate of weight gain (a toxic level). All of the potential pitfalls and ambiguities inherent in acute toxicity testing should be anticipated in subchronic and chronic studies. There are also some unique problems. For example, rats might fail to gain weight appropriately simply because they find the added chemical offensive to their palates. Even rats will draw the line at eating some things, and the investigator obviously cannot ask them how they like it. Few if any rats will refuse the diet altogether, and more commonly they will simply eat less of it than they would of diet not containing the chemical.

The above situation can be recognized because rejection of a diet will result in both a decrease in weight gain and in food consumption. In the case of most true systemic toxic effects, food consumption remains normal, but the animals still fail to gain weight. If the diet is marginally acceptable to the animals, the experiment can still be conducted by resorting to a so-called pair feeding experimental design. Here the amount of treated food consumed by the experimental group on the first day of the experiment is measured, and the control group is rationed to only that amount of food on the second day and so on. At the end, the control group is fed one day longer than the experimental group. Now any observed difference between the control and experimental groups may be ascribed safely to the chemical and not to the amount of food consumed.

The results of feeding young, growing rats, in groups of 10 each, either a control diet or a diet containing, respectively 10, 40, or 160 mg of chemical A per 100 g of feed are summarized in Figure 5–1. Note that both the male and the female rats begin the experiment at the same body weight. As shown in the respective control groups, however, males clearly gain weight more rapidly than females. Thus, it may be inferred that at the start of the experiment the female group was chronologically older than the male group. An equally valid experiment could be conducted by starting male and female rats at the same chronological age but at different mean body weights.

The chemical (Fig. 5–1) was fed for a period of 13 weeks (91 days). For both male and female rats there was no difference in

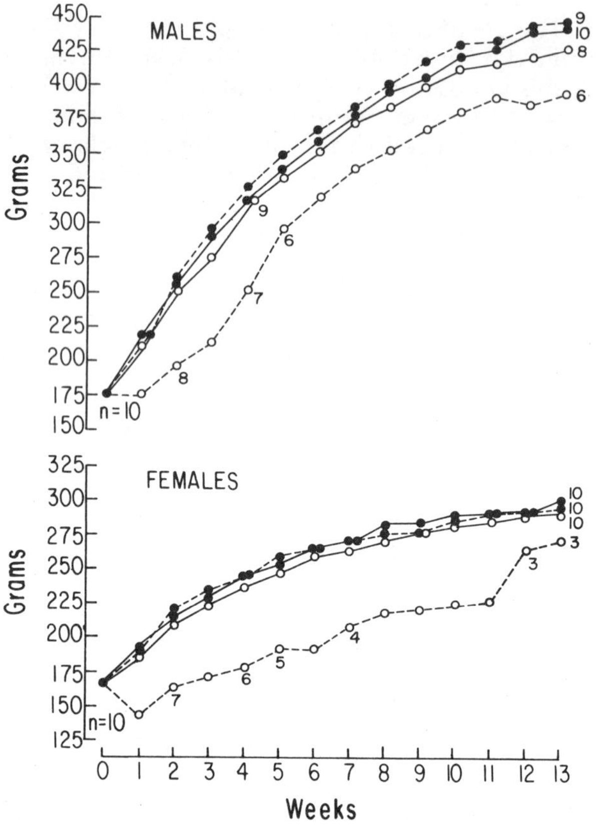

Figure 5–1. Growth curves for rats receiving chemical A mixed in their diets for a period of 13 weeks at levels of (reading in each case from top to bottom) 0, 10, 40, or 160 mg/100 g food. (From Marboom R: Toxicological studies on mebendazole. Toxicol Appl Pharmacol 24:371, 1973.)

weight gain at 10 or at 40 mg chemical A/100 g feed from that in their respective control groups. Therefore, the highest NOEL for both sexes in this experiment was 40 mg A/100 g diet. At a level of 160 mg A/100 g diet the rate of weight gain in both sexes was significantly slower than in their respective control groups. Thus, the lowest toxic level tested in this experiment was 160 mg A/100 g diet, but further testing might reveal lower toxic levels. Because rats consume roughly 10% of their body weight in food each day, it can be calculated that these feeding levels translate into dose rates of about 10, 40, and 160 mg/kg body weight per day, but it would have been more accurate to base that calculation on actual food consumption measurements.

In fact, although not shown in Figure 5–1, food consumption was measured in this experiment. Initially food consumption declined in both males and females at the 160 mg A/kg level, which probably accounts for the sharp drop in body weights during the first week (Fig. 5–1). It is likely that the rats initially found the diet only marginally acceptable; however, by the end of the experiment the total food consumption in all dosed groups was comparable to that in their respective control groups. It may be concluded that the effect seen at the 160 mg A/kg level represents true systemic toxicity.

The small numbers on the growth curves in Figure 5–1 represent the number of survivors in each group at that point in time. All females survived at 10 and 40 mg/kg per day, but 7 out of 10 fed 160 mg per day died during the course of the experiment. The animals were not followed beyond the end of the 91 day period to see if any later deaths occurred. In males one rat died "accidentally" at 10 mg/kg per day, but its death was not believed to be related to the chemical. At 40 mg/kg per day 2 out of 10 males died, and at 160 mg/kg per day 6 out of 10 died. From these observations a number of conclusions may be drawn. The fact that some of the animals died confirms that the chemical exerted true systemic toxicity although it is not necessary to push the feeding to lethal levels to have a successful test result. There is a satisfactory dose-response relationship in that both the slowing in the rate of weight gain and the deaths occurred at the higher feeding levels. There does not appear to be a prominent sex difference with respect to the toxicity of chemical A.

Feeding experiments of the type shown in Figure 5–1 are required for premarketing approval of chemicals for certain uses. The kinds of testing required depend on the intended use for the chemical. Chemicals intended for various uses are regulated by different government agencies. The requirements for a license to manufacture and distribute a therapeutic drug are different from those to manufacture a rodenticide even though the same chemical may be the active ingredient in each. (See also nitroglycerin in Chapter 2.) The use of warfarin as an anticoagulant drug is regulated by the Food and Drug Administration (FDA), whereas the use of warfarin as a rodenticide is regulated by the Environmental Protection Agency (EPA). The types of toxicity testing required differ between the two agencies. The FDA places particular emphasis on testing in humans because that is the species in which the product will be used. In contrast, the EPA places particular emphasis on testing in nontarget species (see below) that may be accidentally exposed even during the proper use of warfarin or other pesticides.

The kinds of testing that might be required by the EPA in order to get a new pesticide on the market are summarized in Tables

5–1 and 5–2. Table 5–1 is a summary of the various kinds of acute toxicity tests that might be required along with their costs in 1972 dollars. The costs were obtained from various contract companies that conduct such testing for a fee. Some manufacturers do their own toxicity testing, but many contract out the work. Also shown for some tests is the time required for their completion.

Under special circumstances the EPA might permit waivers for some of the tests. For example, an inhalation LC50 (Table 5–1) might be waived if it could be proven that it is unreasonable to expect inhalation via the lungs for a given pesticide. The material might be nonvolatile (have a negligible vapor pressure), or the product is not intended for use as a dust or spray. Environmental concerns are reflected by tests in nontarget species, e.g., a fish LC50, an avian LD50 (common test species are the bobwhite quail and mallard duck), and LC50 for crustacae (usual species is shrimp). Remember that the LC50 for fish and crustacae is determined by immersion in aqueous solutions of the pesticide and not by inhalation.

The costs for an oral LD50 include testing in two species selected from mice, rats, guinea pigs, or rabbits. Also included are autopsies of poisoned animals and limited microscopic examination of organs and tissues in an attempt to determine the cause of death. The dermal LD50 is an indication of how readily the chemical penetrates the skin. Skin and eye irritation tests may indicate possible hazards by those routes of exposure.

Initial testing is done with the pure active ingredient in the pesticide product. If the product contains more than one active ingredient, all must be tested so that the costs are doubled for a

TABLE 5–1. COST AND TIME ESTIMATES FOR ACUTE TOXICITY STUDIES ON A PESTICIDE

Test	Cost in Thousands of 1972 Dollars	Time in Weeks
Individual tests		
Oral LD50	1.01	
Dermal LD50	0.65	
Inhalation LC50	0.89	
Eye irritation	0.26	
Skin irritation	0.26	
Avian oral LD50	1.88	
Fish LC50	0.58	
Crustacean LC50	0.62	
Subtotal	6.15	6
Plus formulation	12.30	12
Plus one metabolite	18.45	18

(After Gehring PJ, Rowe VK, McCollister SB: Toxicology: cost/time. Fd Cosmet Toxicol 11:1097, 1973.)

TABLE 5–2. SUMMARY OF COST AND TIME ESTIMATES FOR COMPLETE TOXICITY TESTING FOR A PESTICIDE

Test	Cost in Thousands of 1972 Dollars
Minimum Requirements	
Acute (from Table 5–1)	18.4
Subchronic	100.5
Metabolism, rat and dog	29.8
Reproduction, rat*	35.4
Teratology, rat†	10.4
Chronic feeding, rat and dog‡	155.3
Total	349.8
Additional Tests	
Subchronic	13.2
Teratology, second species§	10.4
Human metabolism	21.1
Mutagenesis	17.2
Human toxicology	50.0
Chronic feeding, metabolite	155.3
Total	275.8

Minimum time required for completion 48 weeks. (From Gehring PJ, Rowe VK, McCollister SB: Toxicology: cost/time. Fd Cosmet Toxicol 11:1097, 1973.)

For comparison one large US contract house has provided quotes effective April, 1990, for the following tests in accord with EPA guidelines:

*Two-generation dietary reproduction study in rats: $220,000.

†Teratology study by oral dosing in rats: $50,000.

‡One-year dietary toxicology study in rats: $200,000, and in dogs: $190,000.

§Teratology study by oral dosing in rabbits: $65,000.

two-ingredient product. The prudent manufacturer might also want to test the two ingredients in combination because they might interact in various biologic systems. The same battery of tests must be performed on the product as formulated for use. The formulated product may contain ingredients that are "inert" from the standpoint of the intended use, e.g., a kerosene base for an insecticide spray product, but these are not necessarily inert in terms of the toxicity of the product to a nontarget species. Many products are marketed in more than one formulation, and each of these must be tested. Testing may also be required for one or more of the major metabolic products, particularly if these turn out to be toxic. Pesticide chemicals may be metabolized to products more toxic than the parent compound. All of this additional testing would increase costs proportionally, and the costs do not include the analytical work that would be necessary to isolate and identify the metabolic products.

The costs for the acute studies appear on the first line in Table 5–2 followed by additional minimal requirements. A subchronic study means a 90-day feeding test as already described in detail. The major metabolites must be identified in both rats and dogs.

A reproductive study in rats often involves three generations to see if the chemical adversely affects mating behavior or numbers of offspring. Teratology studies in rats to see if the chemical produces birth defects when pregnant dams are exposed and chronic feeding experiments in rats and dogs also add enormously to the expense.

Any or all of the following tests might be required including a subchronic test in an additional species and tests for mutagenic activity (see Chapter 20). The time for completion of the minimum requirements is four years and the cost could easily be in the range of 0.5 to 1 million dollars.

Acute, subchronic, and chronic toxicity tests of the type already described can be and often are embellished by various observations in addition to the gross tests of lethality and weight gain. A great deal can be learned from physical examination of the animals. Such signs as ruffled fur, difficulty in walking, lethargy, increased physical activity, and coloration of the tail and paws can be extremely meaningful. Urine may be collected for microscopic examination and various chemical tests. In this case the animals must be housed in special cages that permit the automatic separation and collection of uncontaminated urine and feces. Blood samples may be collected periodically for a variety of chemical tests and for microscopic examination and counting of the numbers of formed elements (red cells, white cells, platelets). Special tests may be used to evaluate how well the liver and kidneys are functioning. At the end of the testing period all survivors are usually sacrificed for microscopic examination of various organs and tissues. The latter are also conducted on animals that die during the course of the experiment.

The special importance of defining the complete dose-response relationship is illustrated in Figure 5–2 for the complicated case of a chemical that is both an essential trace nutrient and, when given in overdose, a toxic metal, namely, copper. Feeding various levels of copper can produce a spectrum of effects that are dose-related, but death is inevitable at both extremes. Because copper is an essential element, if one feeds an artificial diet containing no copper to pigs or sheep for a long enough period of time, the animals will die. The definition of an essential trace nutrient is that it is essential for life. If a little copper is added to the diet, the animals will survive but they will exhibit suboptimal growth and development. This deficiency syndrome will be exhibited in varying degrees of intensity depending on the copper content of the diet and the sensitivity of individual animals.

If more copper is added to the diet, eventually a concentration will be reached that is optimal for most of the animals for normal growth and development (Fig. 5–2). Previously existing signs of deficiency will largely disappear. Further increases in the copper

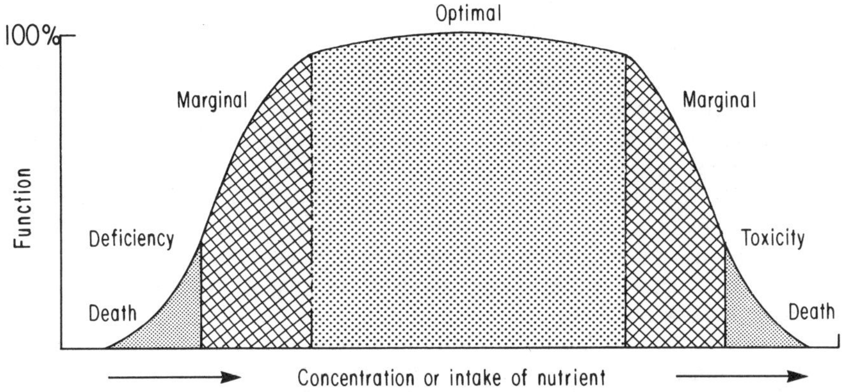

Figure 5–2. The dependence of biologic function on the tissue concentration of an essential trace element as modified by dietary intake. (From Mertz W: The essential trace elements. Science 213:1332, 1981.)

content of the diet, however, will lead to accumulation of the metal in the animals and the expression of toxicity. These signs of toxicity will be different in many ways from the signs of the deficiency syndrome, but some of them may be similar. For example, in both cases growth and development are subnormal. Finally, it is possible to feed copper at a sufficiently high level that the animals will die of copper poisoning.

Figure 5–2 suggests that there is a sharp demarcation in going from one zone of the spectrum to another. Actually, when dealing with groups of animals the lines are less clear because of animal to animal variation. Even with a single animal it may not be obvious when deficiency signs actually begin to appear. In pigs, toxic signs begin to appear when the dietary level reaches 250 ppm. Sheep, however, are much more sensitive, and toxic signs may occur at only 11 ppm, a species difference of 25-fold.

Research over the past 30 years has added seven trace elements (those that are found in living species in amounts measured in milligrams or milligrams per kilogram of tissue) to the list of those already known to be essential in two or more animal species. Oddly, one of these is arsenic. Doubtless further research will expand the list again. The most abundant elements in living matter are carbon, hydrogen, nitrogen, oxygen, and sulfur. These, together with calcium, chlorine, magnesium, potassium, phosphorus, and sodium, are required in grams or fractions of a gram per kilogram per day. Most chemicals of concern in the environment, however, are not essential nutrients, and the concern is only with their NOEL and toxic levels.

SUMMARY POINTS

The 90-day subchronic feeding experiment can be used for most chemicals because it is relatively easy to perform, realistic, and it can define both safe and toxic levels.

The index of toxicity is most commonly a failure of growing animals to gain weight appropriately in comparison with a control group.

A well designed experiment will have at least one level of feeding that has no effect (NOEL), and one or more levels of feeding that significantly retard weight gain.

The kinds of toxicity testing required to market a new drug are specified by the Food and Drug Administration, but new pesticides are regulated by the Environmental Protection Agency even when the products contain the same chemical.

The EPA minimum testing includes acute, subchronic, and chronic tests that can take at least four years and cost hundreds of thousands of dollars to complete.

Chemicals that are both essential nutrients and toxins have a dose spectrum that ranges from death from deprivation to death from toxicity.

BIBLIOGRAPHY

Gehring PJ, Rowe VK, McCollister SB: Toxicology: cost/time. Fd Cosmet Toxicol 11:1097, 1973.

Mertz W: The essential trace elements. Science 213:1332, 1981.

6

CHAPTER

ABSORPTION, DISTRIBUTION, AND EXCRETION OF CHEMICALS

Bill D. Roebuck

As already noted, a chemical can only manifest its toxicity when it can gain access to its target organ, tissue, or cells. Strategies that living organisms have developed to protect themselves from a toxic chemical include mechanisms to prevent its absorption or to prevent its penetration into target organs or cells within the organism. Barring or in addition to those possibilities, the organism may excrete the toxin to prevent its accumulation and the expression of toxicity. During normal life organisms need to absorb and excrete a wide variety of chemicals. For the most part these chemicals represent essential nutrients and metabolic waste products, and some of these are processed in large quantities. Toxic chemicals commonly utilize the same processes of absorption and excretion that nontoxic nutrients and normal waste products use. These processes are controlled by biologic membranes, which divide the body into numerous compartments.

All biologic membranes have common structural features irrespective of whether they are cell membranes or the membranes surrounding various intracellular organelles. This "generic" membrane is found in both embryonic and adult organisms and surrounding the dozens of types of specialized cells of an organism. It serves as a barrier or boundary between the interior of the cell and its exterior environment. As such, it may regulate the uptake of nutrients by the cell and the elimination of its waste products. Toxic chemicals may become inadvertently involved in these processes.

Chemically, biologic membranes are composed of approximately equal parts of protein and lipid (fat) with numerous minor components. Figure 6–1 illustrates in schematic cross-section the features of a simple biologic membrane. The lipid components are largely phospholipids (fats with attached phosphates, $-PO_4$), of which there are several subtypes. The phosphate end of the phospholipid molecule is said to be hydrophilic (attracted to water), whereas the hydrocarbon end of the phospholipid is called hydrophobic (repelled by water). The hydrophilic end of the phospholipid may or may not be electrically charged, but in either case it much prefers to be in an aqueous environment. The hydrophobic region is composed of two fatty acid chains with structures that vary considerably among the various phospholipids.

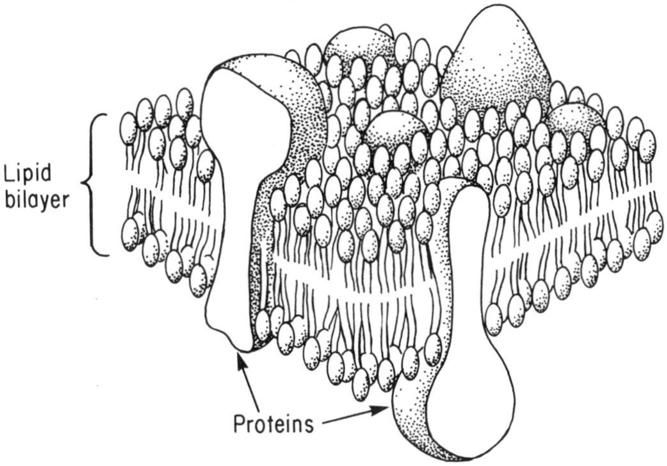

Figure 6–1. A three-dimensional representation of a simple biologic membrane with a lipid bilayer interspersed with transmembrane proteins.

These physicochemical properties ensure that the phospholipids will spontaneously align themselves into a bilayer configuration with the hydrophilic ends projecting into the aqueous media and the hydrophobic tails seeking each other to exclude water (Fig. 6–1). There is even a certain natural tendency for this membrane to form hollow vesicles (micelles) with water both inside and outside the micelle.

Imbedded within and across the lipid bilayer are proteins, either singly or in clusters (Fig. 6–1), which may possess enzymatic activity, act as carriers or transporters of certain molecules across the membrane, or receive signals from the external or internal environment for transmission into or out of the cell. Membranes are dynamic structures in which the phospholipids and proteins move about in the more or less fluid bilayer. Moreover, the cell can adjust the composition of the phospholipids and proteins in order to adapt to changes in the external or internal environment.

The physicochemical properties of individual chemicals determine the rate at which they can cross lipid bilayer membranes as illustrated in Figure 6–2. It is uncommon for molecules to be totally excluded from cells, but some molecules penetrate the membrane so slowly that for practical purposes their concentration inside is zero. The lipid nature of the membrane tends to exclude water-soluble chemicals while favoring the admission of lipid-soluble ones. In general, the smaller the molecule the more rapidly it will cross the membrane. Also as a general rule, molecules that are electrically neutral tend to penetrate membranes more

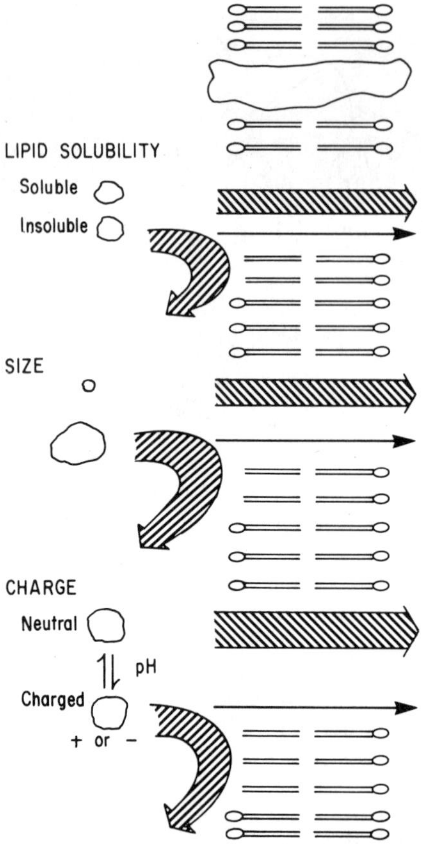

Figure 6–2. The properties that affect the transportation of chemicals across biologic membranes.

rapidly than those bearing electrical charges. These properties are not mutually exclusive and any one molecule may have two or more that work together or that offset each other. For example, a small, neutral, lipid-soluble chemical will penetrate membranes more rapidly than a small, charged, lipid-soluble chemical.

Although some molecules are electrically charged at all values of pH, many organic molecules are called weak electrolytes because the extent to which they exist as charged or neutral forms depends on the pH, a measure of the hydrogen ion concentration or acidity of the solution. At a characteristic pH for each of these molecules, half will be neutral and half will be electrically charged, the so-called pK_a. Moving the pH in one direction from the pK_a will increase the concentration of the charged form, whereas moving in the opposite direction will increase the concentration of the neutral form. By the time one has moved two pH units away the molecule exists almost exclusively in one form or the other (99%). Weak electrolytes may be either weak acids or weak bases. Weak acids tend to be charged at pHs above their pK_as whereas weak bases tend to be charged at pHs below their pK_as. The important point to remember is that the uncharged form tends to cross biologic membranes.

The lipid solubility for a given chemical is expressed in terms of its oil/water (o/w) partition coefficient or ratio. This parameter is easily measured by adding the chemical in question to a vessel containing equal volumes of octanol (instead of "oil") and water. Octanol has chemical properties similar to those of the phospholipids in membranes, and it does not mix with the water layer. The mixture is thoroughly stirred or shaken, sometimes at a controlled temperature. When the octanol and water layers have separated, the quantity or concentration of the chemical is determined in each layer and the ratio is the o/w partition coefficient. Chemicals that tend under these circumstances to concentrate in the octanol layer have a high o/w partition coefficient and high lipid solubility, and, therefore, will in general tend to cross biologic membranes at more rapid rates. The converse is true for chemicals with low o/w partition coefficients.

Several different processes also control the rate at which chemicals move across biologic membranes and enter or exit from cells. The slowest of these processes is simple diffusion, and it is most easily illustrated by molecules in solution or suspended in air. Such molecules are in random motion (Brownian movement). The chemicals move from areas of relatively high to areas of relatively low concentration; their chance "bumping into each other" tends to spread them evenly through the available liquid or air space. Familiar examples are the distribution of a drop of ink in a glass of water or the smell of freshly baked bread in a large room. The driving force for this sort of movement is simply the concentration gradient. In the case of a cell, the chemical would diffuse into the cell from its higher concentration outside until, time permitting, the concentration inside was equal to the concentration outside. The process is called passive diffusion because it occurs spontaneously and requires no energy expenditure by the cell. Cells, however, have other processes for transporting chemicals across their membranes that offer considerably more control over the internal environment of the cell than passive diffusion.

One of these processes is called facilitated diffusion. Like passive diffusion, facilitated diffusion requires no energy output by the cell, and the molecules move spontaneously from an area of high to an area of low concentration. Protein-lined channels within the membrane, however, facilitate this process at least for some molecules while effectively excluding others. Facilitated diffusion is important for the movement of electrically charged, water soluble ions or molecules that would not otherwise readily cross the membrane. If the interior of the water-filled channel is negatively charged, the entrance and passage of positively charged ions or molecules would be facilitated, whereas negatively charged ions would be repelled from the entrance.

Active transport also involves transmembrane proteins, but the expenditure of cellular energy is required to move chemicals across the membrane. This process allows cells to concentrate certain chemicals in their interior even when they are present in low concentrations in the surrounding environment. Presumably the transported chemical is bound to a specialized carrier molecule in the cell membrane, which can be visualized as rotating like a revolving door to release the chemical to the cell interior. Sometimes chemicals are actively transported out at the same time because they bind to another compartment in the revolving door that at the time was facing the interior.

Some large chemicals and even insoluble particles can also be transported into cells by a specialized process called pinocytosis (Fig. 6–3). This process involves a dimpling (invagination) of the cell membrane that deepens to the point where it can be pinched off with the particle inside to form a tiny membrane-enclosed sac within the cell. These steps all require the expenditure of metabolic energy by the cell. If the contents of the sac are to be used by the cell, the sac membrane must be opened and destroyed or recycled. If the sac contents represent a threat to the cell, the sac may move to specialized organelles (lysosomes) that destroy both it and its contents.

Filtration is of particular importance in the formation of urine and the excretion of body wastes or toxic chemicals by the kidney (Fig. 6–4). This process is different from any of the above in that the transport is not into or out of cells, but from one fluid compartment to another. In the kidney, filtration takes place in a series of structures called glomeruli in which pores between indi-

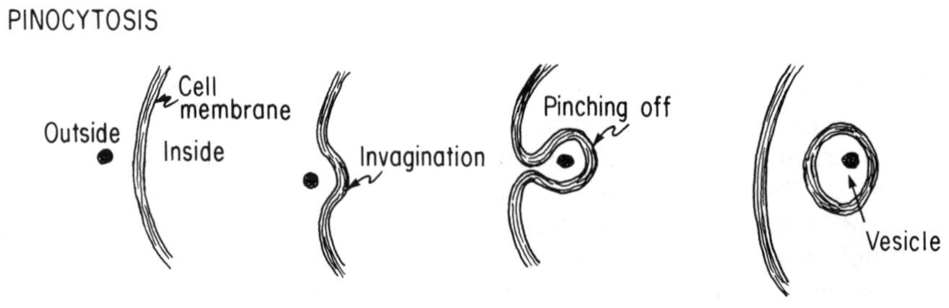

Figure 6–3. The macrotransport process of pinocytosis.

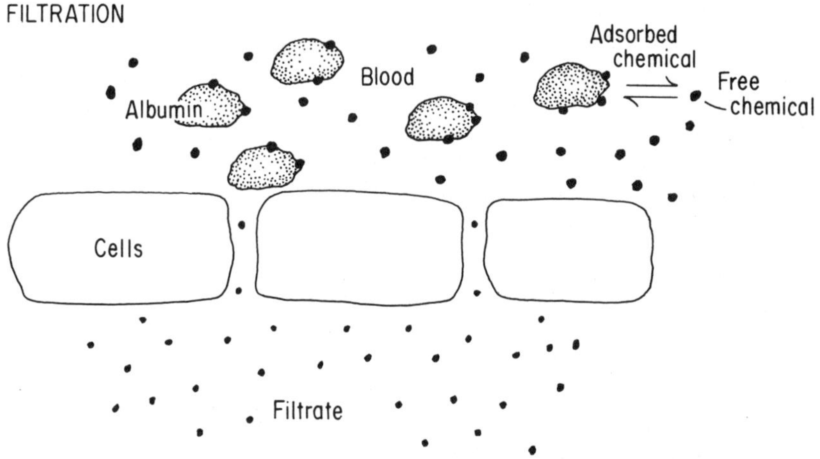

Figure 6–4. The macrotransport process of filtration.

vidual cells separate the blood from the fluid that is to become the urine. The pores are too small for large molecules and various kinds of blood cells to pass, but smaller molecules comprising body wastes or foreign chemicals up to molecular weights of about 60,000 daltons (or gram molecular weight) pass through and may be excreted in the urine.

Transport processes can be influenced by the "free" concentration of the chemical to be transported, and the free concentration of the chemical is influenced by a process called adsorption. Adsorption is the physical binding of a chemical usually to a much larger molecule, structural element, or particle. The bound form of the chemical exists in a state of equilibrium with the unbound or free form. As the free concentration rises the amount bound increases until all the binding sites are occupied. Conversely, the binding sites are emptied by lowering the free concentration. The strength of adsorption and the number of binding sites vary considerably among chemicals. Examples of materials that provide adsorptive binding sites are dietary fibers within the gastrointestinal tract and the protein, albumin, in the blood. The effect of the binding of a drug or chemical to albumin in the blood is to reduce the concentration of the free form which is the form available for transport processes or for exerting toxic or therapeutic effects. Because the binding is reversible, however, the net effect is to slow down any process the free form is involved in, not to bring it to a halt. As the free form is consumed by these processes more of the bound form will be released to take its place. Thus, the bound form constitutes a sort of reservoir of drug or chemical that is slowly released to take part in whatever process is consuming the free form. The binding to plasma proteins also slows the excretion of drugs and chemicals in the urine (Fig. 6–4).

In terms of environmental chemicals, the major routes for absorption are the gastrointestinal tract, the lungs, and through the skin. Scientists working with toxic chemicals, however, may administer them to laboratory animals by other routes, which offer special advantages. Sometimes chemicals are injected through the skin and directly into a large vein. Obviously, this is the most rapid means for increasing the concentration of the chemical in the blood. Because the blood is in equilibrium with all organs in the body, it is also a rapid means of increasing the concentration at target organs or cells. The exact amount of the material given is known, and by definition it is absorbed 100%. Another route that is used commonly for experimental purposes, but almost never for the therapeutic administration of drugs to humans is injection into the peritoneal cavity (the enclosed space that contains the intestines, liver and other organs in the abdomen). The large surface area in the peritoneum has a rich blood supply so that drugs and chemicals are rapidly absorbed. As with administration by mouth and intestinal absorption, chemicals go first to the liver and then, via the general circulation, to the rest of the body. The next slowest route for absorption by injection would be intramuscular, followed by subcutaneous. Important determinants of the rate of absorption include the available surface area, the rate and amount of blood flow through the area, the solvent for the chemical, and the properties of the chemical discussed above (size, solubility, and charge).

When chemicals are inhaled, eaten, or applied to the skin, the amount that is actually absorbed is almost always less than the total amount of chemical to which one was exposed. If the blood level of a particular chemical is followed with time, plotted, and the area under that curve determined for any other route of administration, it can be compared with the area under the curve following intravenous administration. The ratio of the area under the curve for intravenous administration and the area under the curve for the other route is a measure of the amount actually absorbed, or the bioavailability, of the chemical.

The lungs are a common route of exposure to toxic chemicals. Some chemicals damage the lungs themselves without producing systemic toxicity, but others, such as carbon monoxide (see Chapter 1), must be absorbed before systemic toxic effects are expressed. The lungs are structured in such a way as to facilitate efficient and rapid absorption and excretion of essential (oxygen) and waste (carbon dioxide) gases so it is not surprising that gaseous or other airborne chemicals can enter the lungs and in many cases also the blood. The respiratory route can be an extremely rapid way for chemicals to enter the body; in some cases it is second only to intravenous administration. Some occupational dusts are so finely dispersed that they form more or less stable aerosols that can be inhaled and can lead to a variety of chronic diseases such as brown lung (byssinosis in cotton, flax, or hemp workers), black lung (coal miners), and farmer's lung (inhalation of spores from moldy hay).

The architecture of the respiratory system is shown in Figure 6–5. Inhaled air is conducted from the mouth or nose down the trachea and into a branching network of bronchi and bronchioles, which finally terminate in the alveolar region where the major absorptive and excretory processes occur (Fig. 6–5). The total absorptive surface of the alveoli of adult human lungs is approximately that of a football field. A detailed schematic drawing of the surface is also shown in Figure 6–5. The efficiency of gas exchange in the alveoli is increased by two anatomic features. First, the distance between the absorptive surface and the blood that

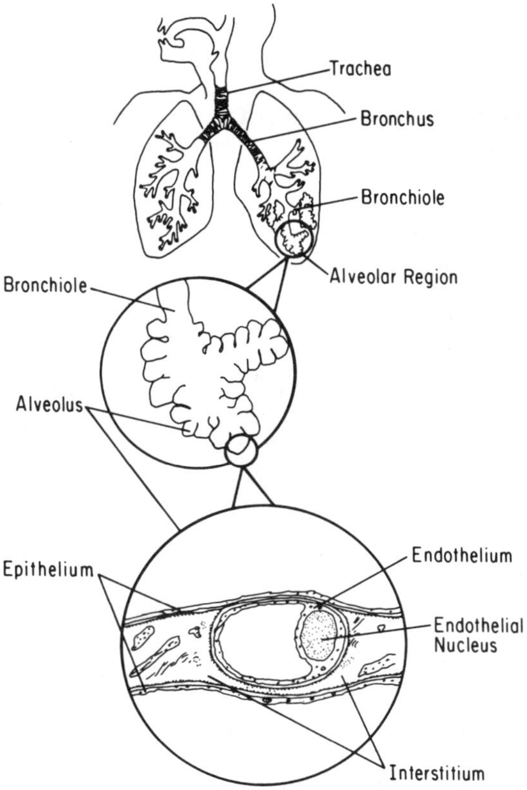

Figure 6–5. The anatomy of the respiratory tract from trachae to alveolus.

must be traversed by gas molecules in the lung is small (1 μm) when compared with the intestine (40 μm) or the skin (100 μm). Second, the rate and volume of blood flow through the lung capillaries is large. Thus, absorbed chemicals are quickly swept away from the absorptive surface by fresh blood. This maximizes the concentration gradient for passive diffusion of inhaled gases from the alveoli into blood, which in turn maximizes the rate of absorption.

The branching arrangement of the bronchi and bronchioles is important in removing particulate matter from the inspired air. As air moves through these passages it must change direction at each point of branching. Particles that are heavier than air resist these changes in direction of flow and may impact on the surface of the passageways before reaching the alveoli. Moreover, at each branch point during descent, the air flow tends to move more slowly, which effects a settling out of suspended particulates. The bronchi and bronchioles, but not the alveoli, are lined with cells that secrete mucus which is a thick, sticky fluid. The same passageways are also lined with ciliated cells, which have tiny, hair-like projections on the cell surface. The cilia are in constant motion, beating to and fro to move a blanket of mucus and any material trapped in it up the passageways and out of the lungs. The mucus blanket is 5 to 10 μm thick, and its rate of movement varies from fractions of a millimeter per minute to several millimeters per minute. This system can remove substantial quantities of foreign particulates in a few hours, but as noted above, it can be overwhelmed so that particles remain in the lungs to produce chronic disease. The mucus can be coughed up and spit out, or it can be swallowed.

Many toxic environmental chemicals exist naturally as gases (carbon monoxide, hydrogen sulfide), but a number of others, although largely existing as liquids or even solids at room temperature, give off variable quantities of vapors, i.e., they are said to have a high vapor pressure. If the chemical has a characteristic odor, this property is readily detected by smelling the air above it, sometimes at considerable distance. Two familiar examples are toluene in model airplane glues and naphthalene in moth balls.

As already indicated in Chapter 4, important factors that determine whether or not toxic signs will occur during or after exposure to a poisonous gas include the concentration of the gas in air, the duration of the exposure and the patterns and rapidity of breathing. Even gases that are biologically inert, such as nitrogen, can be fatal if they are present in concentrations that exlude or lower the normal oxygen concentration. Absorption of toxic gases may occur throughout the respiratory system, but if a gas is readily soluble in mucus, less of it will reach the alveoli where absorption is most rapid. This scrubbing action together with the constant renewal and removal of mucus serves to protect the lungs and the organism from systemic toxicity. Some gases may react chemically with mucus and effectively be detoxified. The nasopharyngeal area (Fig. 6–5) of the respiratory tract is particularly effective in scrubbing inhaled air, but breathing through the mouth as in heavy exercise effectively bypasses this protective mechanism. The more lipid-soluble the gas, the less likely it is to react with the mucus blanket. Such gases are rapidly absorbed in the alveoli and tend to produce systemic poisonings.

Aerosols may consist not only of solid particulates but also of finely dispersed mists of liquid droplets. Vapors, fumes or gases may be adsorbed to the surface of aerosol particulates or dissolved in liquid droplets. In these cases the toxic effects may be caused by the adsorbed or dissolved chemicals instead of the carrier aerosol. In other cases the toxicity may result from synergistic effects between the two.

The size of the aerosol particles critically determines where deposition will occur in the respiratory tract. In any given aerosol a large range may exist in the sizes of the particles, in their weights, in their shapes, and surface areas. Among aerosols one must expect differences in chemical composition and in solubilities. These factors and the ways in which they influence pulmonary deposition are shown in Table 6–1. The larger solid and liquid particles never reach the alveoli; they are intercepted by nasal hairs or impact into the mucus blanket. The smaller particles that do reach the alveoli are cleared much more slowly. They may slowly dissolve and be absorbed, or they may be phagocytized by lung macrophages by the process of pinocytosis. The macrophages may leave the lung by way of the lymphatic system.

TABLE 6–1. GENERAL FACTORS INFLUENCING THE DEPOSITION OF AEROSOL PARTICLES IN DIFFERENT AREAS OF THE RESPIRATORY TRACT

Region of Deposition	Particle Size (μm)*	Deposition Process	Removal Process
Nasopharyngeal	5	Interception, impaction	Cough or sneeze
Trachea and Bronchi	2–5	Impaction, sedimentation	Mucociliary clearance
Bronchioles	1–2	As above	As above
Alveoli	<1	Diffusion	Phagocytosis, dissolution

*One μm = 0.0001 cm or about 0.0002 in.

Inhaled chemicals may produce any of three types of diseases in the lungs: pneumoconioses, allergic reactions, or cancer. Pneumoconiosis is a general term applied to the production of fibrous tissue in the lung, usually in response to a series of chemical insults. In this condition the available surface area for gas exchange may be reduced, and it can even interfere with the mechanics of breathing. Black lung was mentioned earlier. Certain mineral fibers, such as asbestos, may cause not only a pneumoconiosis, but also a specific form of cancer. Byssinosis and farmer's lung probably have an allergic basis, as also does the severe reaction some people exhibit on exposure to formaldehyde vapors. The incidence of lung cancer could be enormously reduced by abstention from cigarette smoking, but other chemicals are also recognized to cause lung cancer.

The gastrointestinal tract is made up of several separate but interrelated organs (Fig. 6–6). It is a continuous tube from the mouth to the anus in which the contents are actually external to the body, but they are in intimate contact with a huge absorptive surface.

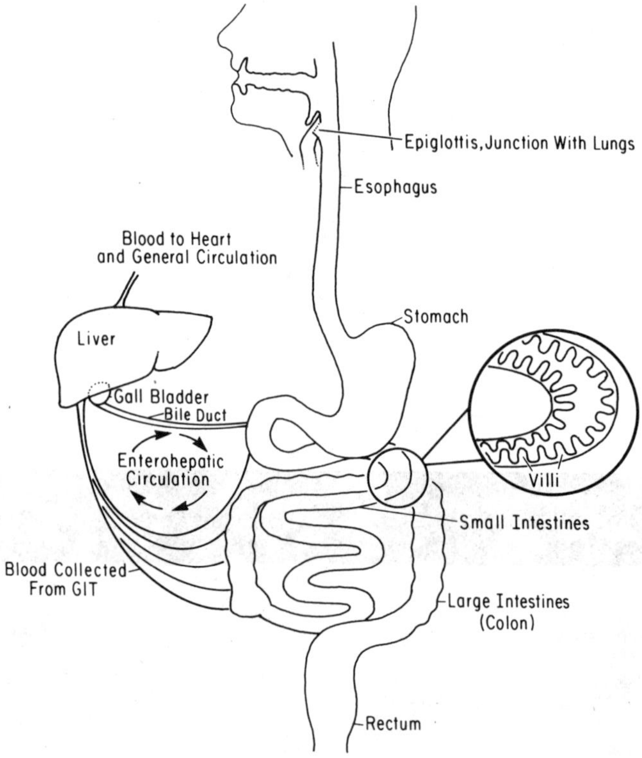

Figure 6–6. The anatomy of the gastrointestinal tract.

Although absorption can occur along the entire length of the tract, the bulk of absorption occurs in a specific and relatively short segment. The gastroinestinal tract does provide for efficient absorption of nutrients needed by the body. Food is mechanically reduced to small conglomerates by chewing, and then it is chemically digested to a well mixed slurry of relatively small molecules. Extensive mixing continues throughout most of the intestines, and the chemical composition of the mixture is modified by adjustments in pH and the addition of enzymes. These changes facilitate the absorption of specific nutrients, but unfortunately, they often facilitate the absorption of foreign chemicals as well.

Absorption can occur in the oral cavity, e.g., the drug nitroglycerin is placed under the tongue to relieve the chest pain arising from an attack of angina, but this is not in general a quantitatively important route. The chewing and wetting of food in the mouth, however, is an important aspect of mixing to facilitate digestion. The epiglottis (Fig. 6–6) is found at the junction of the gastrointestinal and respiratory tracts. If it is not closed during swallowing, foreign material may get into the lungs, perhaps with catastrophic results. Conversely, fluids expelled by the lungs may be swallowed or expectorated. In addition to digestion, the stomach serves as a reservoir for the food mass which is slowly released into the intestine. In response to its content of food and water special cells in the stomach secrete hydrochloric acid as an aid to digestion. Acidification of the stomach contents may modify the toxicity of foreign chemicals. A few chemicals are actually destroyed by the high acidity, but this process cannot be relied upon as protection against most foreign chemicals. While high acidity may hasten the absorption of some chemicals which are weak acids, it can slow the absorption of weak bases.

Most absorption occurs within the first few feet of the small intestine; so absorption increases rapidly after food leaves the stomach. It is here that the bulk of the amino acids, sugars, fats, vitamins and minerals are absorbed. As already indicated, mixing continues, and the acid added in the stomach is largely neutralized. Bile acids are added as an aid to the absorption of fat and lipid-soluble chemicals just as a detergent helps in removing grease from cooking utensils by making it miscible with water (see Chapter 15). The cells lining the small intestine have finger-like projections called villi and are arranged in such a way as to maximize the absorptive surface area. The interior surface area of an adult small intestine is about 1800 ft^2 although its length is only 9 ft and its diameter is less than 2 in. Absorption is also enhanced by the rich blood flow from the intestines, which is further increased by the presence of food. Absorption can occur by any of the processes discussed above.

Some chemicals can become involved in a process called enterohepatic recycling, which is capable of greatly prolonging their stay in the body. The blood from the gastrointestinal tract passes through the liver before it returns to the general circulation. This blood contains high concentrations of absorbed nutrients and any foreign chemicals taken by mouth. The liver can act enzymatically on either of these (see Chapter 7). In some cases the toxic chemical or its metabolic product are then excreted via the bile back into the intestine. From the intestine it may be reabsorbed to repeat the enterohepatic cycle a number of times, and this process can greatly slow its eventual excretion.

Enterohepatic recycling can be thwarted if material is present in the intestine to which the recycled chemical can bind. Vegetable fibers were mentioned above as adsorbtive substances. This principle is exploited in clinical toxicology in two ways through the use of activated charcoal, a material with an enormous adsorptive surface to which a wide variety of chemicals bind strongly. If a toxic chemical, which undergoes intensive enterohepatic recycling, has already been absorbed, activated charcoal, which does not leave the gastrointestinal tract, can be given to prevent the reabsorbtive phase of the cycle. If the toxic chemical is still in the stomach, activated charcoal can be given to help prevent its initial absorption. If the charcoal is removed as by vomiting or a "stomach pump" together with its absorbed material, the process is more efficient, but some antidotal effect is often observed even when there is no attempt to remove the charcoal.

Nonpathogenic bacteria live normally in the lower portions of our intestinal tracts. These bacteria are capable of acting on normal nutrients and some foreign chemicals. In a few cases these chemical changes are beneficial whereas in others they are harmful. The bacteria contain enzymes that are not found in normal animal cells, so that some chemicals can undergo unique biochemical changes in the lumen of the intestine that cannot occur anywhere else in the body.

In summary, a major role for the gastrointestinal tract is to provide for the efficient absorption of essential nutrients. The large surface area of the interior of the intestine and its plentiful blood supply contribute to this essential function. Unfortunately, it cannot at the same time exclude completely the passive absorption of toxic chemicals taken by mouth. In a few cases toxic chemicals are actively absorbed because their chemical structures resemble that of natural nutrients. The rate at which a chemical is absorbed can often mean the difference between no expression of systemic toxicity and death caused by poisoning. Many variables can influence absorption, including the rate at which the stomach is emptied, the extent of mixing of the food mass, the pH, which influences the degree of ionization of some chemicals, and the rate at which solid chemicals dissolve in water.

The skin is another potential route for the systemic absorption of toxic chemicals. The skin is different from the respiratory or gastrointestinal tracts because one of its primary functions is to prevent chemicals from entering or leaving the body whereas the lungs and the gut are primary portals of entry and exit. This functional difference is reflected in the fact that relatively few chemicals penetrate through the intact skin in amounts sufficient to produce systemic poisonings.

Figure 6–7 illustrates the important structural features of a typical cross-section of human skin. The two major layers are the epidermis and the dermis. The outer layer of the skin (epidermis) is subdivided into two layers, the stratum corneum and the stratum germinativum. The stratum germinativum is composed of germ cells that continually divide to provide new cells for the stratum corneum, although these new cells are destined to die. The dead cells of the stratum corneum provide a tough layer on the outermost surface of the body that can withstand the drying action of air, retard the loss of body water, and resist friction and abrasion. They also retard the absorption of water and lipid-soluble agents from the environment, which would have to cross

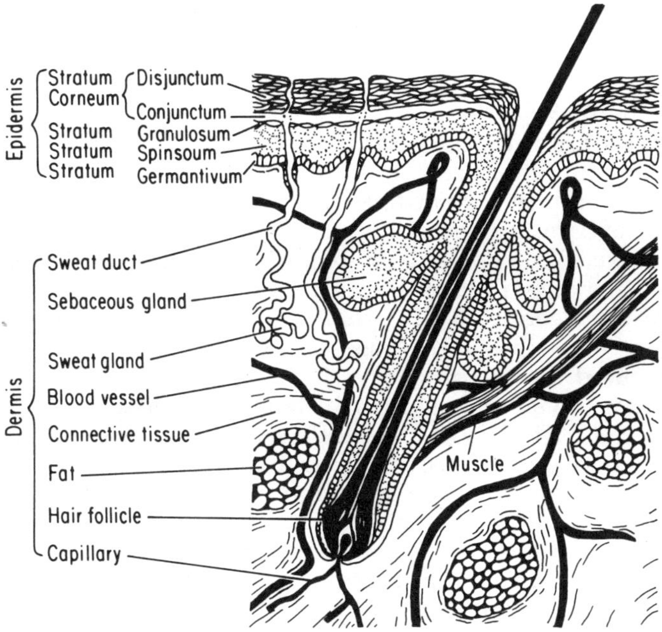

Figure 6–7. The organization of the skin as a biologic barrier.

several layers of dead cells and membranes of considerably greater thickness than other membranes in the body.

Although the skin is not a perfect barrier, it does retard the absorption of most chemicals and completely excludes some. Because the outer layer of cells are dead, they have no active absorptive processes, and all absorption through them is by passive diffusion. Removal of or damage to the outer layer, however, would allow chemicals to come into contact with living cells with active transport capability and close proximity to blood supplies. The deeper regions of the skin contain blood vessels and the cells involved in the production of sweat, oils, and hair. The skin has many minute pores through the epidermis (Fig. 6–7) for sweat and oil glands and the shafts of hair, but *in toto* these constitute only a small fraction of the total surface area of the skin. As such these pores are not an important route of entry for chemicals.

In general, chemicals are excreted from cells by the reverse of the same processes that accounted for their absorption. Again, lipid solubility and concentration gradients are key factors in passive diffusion. Adsorption to intracellular proteins or other macromolecules can play an important role. A new aspect, however, is the fact that the chemical may have undergone biotransformation while in the cell so that excretion is not simply the reverse of absorption. Chemicals can be eliminated from the body in exhaled air or any fluid, but the most important routes are the urine and feces. The feces will contain residual amounts of chemicals that were not completely absorbed or were absorbed and then excreted with the bile.

Exhaled air is important for the excretion of carbon monoxide and a few other chemicals, but it can be used for the noninvasive monitoring of some chemicals in blood. The Breathalyzer or Intoximeter is commonly used as a measure of blood alcohol levels (see Chapter 19). A chemical in blood with a high vapor pressure, such as ethanol, will exist in a state of equilibrium with alveolar air. As the concentration of ethanol in blood rises, its concentration in alveolar air will increase proportionately and be detected in the expired air.

The biliary system has already been described as a potential route for the excretion of chemicals (Fig. 6–5). The liver occupies a key position in relation to the gastrointestinal tract because all the venous blood draining from the intestines passes through it before joining the systemic circulation. One important function of the liver is to chemically biotransform absorbed foreign chemicals along with specific nutrients (see Chapter 7). Some of these are excreted in the bile and eliminated in the feces. Each day the liver produces 0.5 to 1.0 liter of bile, of which about 80% is recycled in the enterohepatic system. Some liver cells have the capacity to transport actively chemicals into the bile whereas others enter bile by passive diffusion. Few generalities can be offered about the types of chemicals that are excreted in bile because the system has not been intensively studied, but it can be affected by several factors. Liver injury tends, in general, to interfere with this pathway. Bilirubin, a waste product of hemoglobin metabolism, is excreted in the bile. In some types of liver injury bilirubin excretion is inhibited, and it accumulates in blood to result in jaundice. In other cases chronic exposure to certain chemicals enhances bile production and flow leading to an enhancement in the excretion of chemicals by the biliary route.

Urine is probably the major route for the excretion of toxic chemicals and their biotransformation products. Urine formation by the kidneys is a highly complex process, which has already been discussed in part above. The kidneys receive a substantial fraction of the total blood flow and the plasma is subjected to filtration in the glomeruli. The free fraction of chemicals, i.e., that fraction not bound to plasma proteins will be cleared in this process, but the kidney also permits passive diffusion of chemicals into urine and in some cases it participates in active transport. In general, urine is slightly acidic relative to the blood so that some weak bases are excreted more rapidly than weak acids, but the urine pH can be increased simply by giving baking soda (sodium bicarbonate) so that the converse becomes true. This procedure is commonly used to hasten the excretion of aspirin from poisoned patients. Lipid soluble chemicals can be passively reabsorbed from the urine as the kidney attempts to conserve body water by concentrating the urine fluid volume. In some cases this greatly reduces the net urinary excretion of chemicals. If the urine volume is low, chemicals in urine may be concentrated to the point that they precipitate out of solution and physically obstruct urine flow.

Other minor natural routes for the excretion of chemicals include mammalian breast milk. A few lipid soluble drugs and chemicals are excreted in breast milk. Although this may represent only a small fraction of the total excreted chemical, breast milk may constitute the sole source of food intake for infants. Hence, infants may be exposed to relatively large amounts of the chemical and exhibit signs of toxicity. The excretion of chemicals into seminal fluid may result in infertility or even sterility (see Chapter 20). Very small amounts of specific chemicals may be found in hair, nails, sweat, and saliva. Again, these may serve as noninvasive means of monitoring exposure or body burdens. Hemodialysis with an artificial kidney is an efficient way of increasing the filtration of the blood to remove natural waste products or absorbed chemicals, but it is obviously not a natural route of excretion.

SUMMARY POINTS

Chemicals must cross many lipid biologic membranes during their absorption, distribution, and excretion from the body.

The chemicals that most readily cross biological membranes tend to be more lipid soluble, smaller, and without an electrical charge.

The processes by which chemicals cross biologic membranes range from the slowest, diffusion, to the most rapid and complex, active transport.

Chemicals can bind reversibly to many natural sites in the body, but only the free form of the chemical has biologic activity and can be excreted.

The major routes for absorption of environmental chemicals are the gastrointestinal tract (after swallowing), the lungs (after inhalation) and the skin (after contact).

The major route for excretion of chemicals is in urine, but in special cases the feces, the lungs, and breast milk are routes for excretion.

BIBLIOGRAPHY

Alberts B, Bray D, Lewis J: Molecular Biology of the Cell, 2nd ed. New York, Garland Publishing, 1989.

Hopkins CR: Structure and Function of Cells. London, WB Saunders, 1978.

Klaassen C, Amdur MO, Doull J: Cassarett and Doull's Toxicology: The Basic Science of Poisons, 4th ed. New York, Macmillan, 1991.

Levine RR: Pharmacology: Drug Actions and Reactions. Boston, Little, Brown and Co, 1983.

7

CHAPTER

BIOTRANSFORMATION

The biologic effects of chemicals are often terminated long before the chemicals are eliminated from the body. Although exceptions are recognized, drugs or chemicals are not usually excreted in the urine or feces in the same form in which they were absorbed. They commonly undergo chemical changes in a process referred to as biotransformation (also less precisely as metabolism), and the biotransformation products tend to be less biologically active or totally inactive in comparison with the form that was absorbed. Thus, when an anesthetic dose of a sedative drug is given and the person falls asleep, he or she wakes up because the drug has been converted to an inactive form by biotransformation, not because of excretion. Eventually, however, the inactive biotransformation products are eliminated by excretion. A few drugs and chemicals do not undergo biotransformation. These drugs have one thing in common in that their effects are unusually long-lasting in comparison with other drugs having the same effect, but which undergo biotransformation. Clearly, biotransformation is a major determinant of the duration of action of absorbed drugs or chemicals so it can represent a defense mechanism against toxicity. Unfortunately, biotransformation reactions can be a double-edged sword. Although it is less common, some drugs and chemicals are activated or made more toxic by biotransformation. In such cases instead of biotransformation being exclusively a defense mechanism, it can work to our detriment.

The number of different types of biotransformation reactions is astonishing, but the largest group are oxidative in nature, i.e., an oxygen atom is introduced into the molecule. Chemicals may undergo oxidation in many ways, and the products are usually more water-soluble than the starting molecules so they are more readily excreted in the urine. Indeed, if this is the primary purpose for biotransformation one can rationalize why some chemicals are activated instead of inactivated. Biologically active xenobiotics tend to be more lipid-soluble and less water-soluble than the materials that the body uses to derive metabolic energy. If the machinery for carrying out biotransformations was encased in lipid membranes it would permit access of the xenobiotics and exclude normal carbohydrates, amino acids, and other materials used in energy metabolism. This machinery would have no way of recognizing a toxic from a nontoxic xenobiotic, so that in its mission of turning lipid-soluble xenobiotics into more water-soluble biotransformation products to facilitate excretion in the urine, it might "accidentally" make a more toxic product.

The major site for biotransformations in the body is the liver. Although many other tissues can also carry out some reactions, the liver has the largest repertoire and handles by far the largest volume. The enzymatic activity was first localized by successive centrifugations of liver homogenates where the various fractions separated represent different subcellular organelles. The enzymes were most clearly associated with the microsomal fraction, and they are still often called hepatic (for liver) microsomal enzymes. Microsomes are isolated as lipid-covered beads, but they originate from a complicated subcellular structure called the endoplasmic reticulum (ER). There are two types of ER, the smooth and the rough. The rough ER carries out the synthesis of proteins needed by the cell including the enzymes for biotransformation, but the enzymes are transported so that the bulk of the biotransformation activity is associated with the smooth ER. The ER is a labyrinth-like structure superficially resembling the layers of an onion with access pathways between the layers and the cell nucleus. When the cell is chopped into pieces by homogenization, the ER is also chopped up into the lipid-encased beads or microsomes that are isolated by centrifugation. The biotransformation enzymes are located in the water-filled interior of the beads or rods.

Most enzymatic reactions are highly specific for only one substrate or a small series of closely related chemicals. One distinguishing feature of enzymes carrying out biotransformation reactions is their lack of specificity. Thus, they are prepared to act on newly synthesized xenobiotics to which humans have never before been exposed. This is possible because of two other unique features of the system. At first it was believed that only one or possibly two enzymes were involved. The first one to be discovered was named cytochrome P450 because of the position of its absorption band when it is bound to carbon monoxide. Now it is known that P450 is not one or two enzymes, but a family of closely related enzymes or, as they are called, isozymes. New isozymes are still being discovered and characterized so there may be dozens of different enzymes, each with a somewhat different spectrum of substrates on which it can act. It is the multiplicity of enzymes that accounts in part for the apparent lack of specificity of the system, but even single isozymes of P450 can often carry out more than one type of biotransformation reaction.

The other unique feature of the system is that the enzymes are inducible. Liver cells have the ability to make new enzymes on demand, and to add them to the supply already in existence so that the capacity to carry out biotransformations is greatly increased. The logical signal to start induction is simply increased exposure to xenobiotics. Not all xenobiotics are equally effective inducers, but the best inducers seem to be chemicals that are lipid-soluble and that have a relatively long half-life in the body. Many inducers are themselves substrates for the isozymes, but that is not an absolute requirement. The most potent inducing agent yet discovered is dioxin (TCDD) (see Chapter 11), but other examples include the DDT-type insecticides (see Chapter 10), an antiepileptic drug (phenobarbital), and polycyclic, carcinogenic hydrocarbons such as benzpyrene. Benzpyrene is found in tobacco smoke, and smokers are usually exposed to levels above the nonsmoking population. It is also found in high concentrations in charcoal-broiled meat. Ethanol, and other constituents of alcoholic beverages, are also effective inducers.

Different chemicals induce different spectrums of isozymes, which contributes to the illusion of the nonspecific nature of biotransformation reactions. If a particular chemical is inactivated by biotransformation, induction increases the protection against toxicity, but if the chemical is activated by biotransformation, induction results in increased toxicity. When the exposure to an inducing agent is terminated, the synthesis of new enzymes slows down and the levels return to the pre-exposure state. Maximal induction may occur within 24 hours or may take up to several days. Levels return to pre-exposure activity over days to a period of a few weeks. Induction occurs in all tissues that contain P450 isozymes, but because the liver has the richest supply, it responds to induction to a greater degree. Commonly, inducing agents increase the synthesis of several isozymes so that induction is no more specific than biotransformation itself.

Most enzymes can be inhibited, and the P450 isozymes are no exception. As the concentration of chemical being biotransformed in the liver increases, a point is reached at which all available enzyme is saturated, and the rate of biotransformation reaches a maximum that cannot be exceeded. If a second chemical that is acted upon by the same isozyme is then added, its rate of biotransformation is much slower than in the absence of the first chemical. Because any chemical that is biotransformed by P450 isozymes is a potential inducer as well as a potential inhibitor, and because chemicals can either be inactivated or activated by biotransformation, the door is opened to a legion of possible chemical-chemical interactions that are time- and concentration-dependent. Because of the necessity for one chemical to reach saturating concentrations in the liver, P450 inhibition is less common than induction, but some chemicals are widely known to be effective inhibitors at readily achievable concentrations, including ethanol and the antiulcer medication, cimetidine. If chemical A is activated by biotransformation, prior exposure to chemical B may have resulted in induction and increased toxicity because of chemical A. If, however, chemicals A and B are present at the same time, B may inhibit the biotransformation of A, resulting in decreased toxicity.

These chemical interactions are largely unpredictable, but can be discerned by experiments with laboratory animals. It is common for patients to be taking more than one drug at a time for a given illness as well as self-prescribed remedies and recreational drugs. Thus, a large body of clinical experience adds to the animal tests, and certain patterns of dangerous drug-drug interactions based on their biotransformation are now well known and avoided. The introduction of new drugs, however, is sometimes accompanied by the recognition of a new drug-drug interaction. It is extremely fortunate that so few of these have important clinical significance.

Biotransformations often occur in two steps. The activity of P450 isozymes is referred to as a Phase 1 reaction. Phase 1 reactions not only result in more water-soluble products, but they also often put a "handle" on the molecule such as introducing a hydroxyl (-OH) group in the place of a simple hydrogen. Most of the subsequent Phase 2 reactions do not take place in the ER, but in the cell cytosol, which has an abundant supply of water-soluble metabolic intermediates, such as sugars, amino acids, and inorganic anions like sulfate. Phase 2 reactions are also called synthetic reactions or conjugation reactions because they involve the joining together of the product of a Phase 1 reaction and a naturally occuring metabolic intermediate such as glucose, glycine, or sulfate. This bonding occurs via the handle introduced in Phase 1, and it results in products that are even more water soluble and even more readily excreted in the urine than the products of Phase 1. Again, however, this two phase pathway is only a generality. Some xenobiotics undergo much more complicated and extensive biotransformation than others. Dozens of biotransformation products may arise from some chemicals.

Many factors converge to influence the rate of biotransformation reactions including the general state of health of the liver. Liver diseases, such as cirrhosis or cancer, may result in impairment or slowing of rates of biotransformations and the ability of chemicals to induce the synthesis of P450 isozymes. The doses of some drugs may have to be reduced in patients with impaired liver function to avoid toxicity. Neither humans nor animals are born with a full complement of P450 isozymes. It may take several years for human infants to acquire adult levels of biotransformation activity. For this reason the doses of some drugs in infants may have to be adjusted in ways that are out of proportion to the difference in body weight alone. Less is known about the geriatric population in the absence of liver disease. It certainly deserves more study because this is the population that takes the most drugs. Wildly different rates of biotransformation reactions may be encountered among different species. Indeed, this is the major factor accounting for species differences in toxicity. Strain and sex differences are recognized as well, but sex differences among humans are usually not clinically important.

SUMMARY POINTS

The biologic activity or toxicity of most chemicals is terminated by chemical reactions (biotransformations) long before the chemicals are finally excreted from the body.

Less commonly biotransformation can result in a more active or more toxic product, but in either case the product is more readily excreted in the urine.

The liver is the most important site for biotransformations; the enzymes are found in the smooth endoplasmic reticulum of liver cells and are isolated as microsomes.

The first enzyme found was named cytochrome P450; it is now known that this is a family of isozymes, none of which can carry out the full spectrum of biotransformations.

Various inhibitors are known for biotransformations and liver damage slows the reactions; the result may be a slowing of detoxification or a failure of activation.

The system is highly inducible both by its substrates and by other chemicals, which suggests an autoprotective role, but only against chemicals that undergo inactivation.

BIBLIOGRAPHY

Lippmann M: Toxic chemical exposure and dose to target tissues. In Lave LB, Upton AC (eds): Toxic Chemicals, Health and the Environment. Baltimore, Johns Hopkins Press, p. 114, 1987.

Kamrin MA: Human variability and toxicity: the human body and toxic chemicals. In Kamrin MA (ed): Toxicology. Chelsea, MI, Lewis Publishers, p. 19, 1988.

Lu FC: Absorption, distribution and excretion of toxicants. In Lu FC (ed): Basic Toxicology. New York, Hemisphere Publishing, p. 9, 1985.

Pines, M: Medicines and You. Washington DC, US Department of Health and Human Services, Public Health Service, National Institutes of Health, Publ. No. 81-2104, 1981.

8

CHAPTER

FOOD AND PESTICIDE LAWS: SETTING TOLERANCES

Laws for the regulation of the pharmaceutical, food, and pesticide industries have evolved slowly in the United States, and they have many similarities. The Federal Insecticide, Fungicide, and Rodenticide Act (FIFRA) of 1947 was the original legislation intended to protect the public and the environment from pesticides. This original law designated the Department of Agriculture as the regulatory agency. In 1972 FIFRA was comprehensively revised as the Federal Environmental Pesticide Control Act (FEPCA). The new law designated the EPA as the regulatory body so that pesticides could be regulated along with other air and water pollutants, solid waste, noise, and radiation. The EPA itself had been previously created by the National Environmental Protection Act of 1969.

The pesticides covered under the act include all substances other than antibiotic drugs intended for destroying or repelling any pest organism ranging from viruses and bacteria to rodents, and any substance intended to destroy or regulate plant growth. The law specifically excluded new veterinary drugs and animal feeds.

Two mechanisms are used by the EPA to control the movement of pesticides in interstate commerce. The original device, like that for foods, drugs, and cosmetics (see below) was the label. Each pesticide product was required to bear an EPA-approved label. The labelling provisions are still in effect, but emphasis has gradually shifted to the registration of the product with the EPA. A valid registration number must appear on the label of each product; otherwise, the sale, distribution, advertisement, shipment, or delivery of that product from one state to another is prohibited by the law. The establishment producing the product also must be registered with the EPA.

A complete application for the registration of a pesticide can be quite voluminous because of the extensive toxicity data required. In addition, the application must include the name of the pesticide and the product, the complete chemical formula(s) for all active ingredients and a copy of the proposed label. The EPA administrator is required to publish periodic guidelines and revisions of procedures and data required in support of the registration. If, after study of the toxicity data, the EPA concludes that the pesticide can be used safely, the next step is to approve the label.

Labels on all consumer products contain important information, and because of their potential toxicity, this is particularly true for pesticides. The average American consumer probably does not spend enough time reading them to extract the information that is provided for his own protection. An adequate label for a pesticide must contain the name of the product and the manufacturer, the net contents of the package, a complete list of all active ingredients, a valid registration number, directions for use, which, if complied with, are adequate to protect the public and the environment. The label may also contain warning or cautionary statements such as "avoid contact with the skin" or "avoid breathing vapors."

The EPA administrator is empowered to classify a registered product for general use if it will not cause "unreasonable adverse effects in the environment," or for restricted use if it may cause such effects in the absence of regulatory restrictions. The administrator can grant a so-called experimental use permit if he determines that the applicant needs such a permit in order to accumulate the data needed for full registration. He may also set a temporary tolerance level for the amount of pesticide residue permitted on crops at harvest if there is reason to believe that the experimental use will result in contamination of crops used for human food or livestock feed. All registrations are automatically cancelled after 5 years unless the registrant actively pursues reregistration.

The administrator is further empowered to issue a notice of intent either to cancel a registration, to change its classification, or to hold a hearing to determine whether or not such actions should be taken. If the administrator determines that suspension of a registration is necessary to prevent an "imminent hazard to human health" such an action can be effective and final 30 days after notification of the manufacturer.

The application for the registration of an establishment to manufacture or formulate pesticides must include data on the types and amounts of all pesticides currently being produced, sold, or distributed by that establishment. This information is a valuable resource for all segments of our society. Other provisions of FEPCA, which serve to protect the public and the environment, make it unlawful to detach, alter, or destroy the label on a pesticide product. Also prohibited is the use of any product in a manner inconsistent with its labelling, and the addition of any substance or the removal of any substance from a pesticide product. In consultation with other relevant agencies the EPA is required to establish regulations and procedures for the safe disposal or storage of empty packages or containers for the pesticide. The EPA must also accept for safe disposal at locations convenient for the manufacturer any remaining pesticide after its registration has elapsed or has been cancelled, if such a request is made by the establishment.

This latter "indemnification" rule created an enormous financial burden for the EPA. It has banned four pesticides on an emergency basis: ethylene dibromide (EDB), dinoseb, 2,4,5-T, and Silvex. Their storage and disposal could cost as much as 2 to 3 times the EPA's pesticides program budget. Because by law the agency must shoulder these costs, money must be diverted from regulatory activities to pay for it. This rule is obviously an important insurance policy for the pesticides industry, but it places the EPA in the anomalous position of paying for its own decisions to protect the public and the environment. This indemnification rule was struck down in return for an extension of the patent rights on pesticides to manufacturers to cover the higher costs of safety testing and development. An escalation of the fee for registration will bring new revenues into the agency to speed up the review of old pesticide products, and the responsibility for storage and disposal of banned products has been shifted to the manufacturer.

Practical and legal issues involving pesticides and foods often overlap. Laws regulating the purity of American foods and drugs are less than a century old. Congress passed the Pure Food and Drug Law and the Meat Inspection Act on the same day in 1906. The Department of Agriculture was designated as the regulatory body for animal meats, whereas the Department of the Interior was made responsible for such aquatic foods as fish and shellfish. These laws also used the label as the regulatory device to prohibit the interstate commerce of any misbranded or adulterated foods, beverages, or drugs. It is said that Theodore Roosevelt, then president, took a personal interest in these laws and greatly expedited their passage, particularly the Meat Inspection Act, after reading a book by Upton Sinclair called *The Jungle.* Others who had read it at that time may well have felt the same way.

Sinclair was rightfully concerned about many of the social problems in turn-of-the-century America. *The Jungle* was intended as a polemic on a wide variety of social ills as seen through the eyes of recent emigrants from Europe. Perhaps the title gave rise to the expression, "It's a jungle out there," in reference to urban life. The book was set in and around the Chicago stockyards, which at that time was the hub of the American meat packing industry. Perhaps only as background color, however, *Jungle* contained sufficient revolting details about the filthy and draconian working conditions and unsanitary procedures (many of which were probably authentic for the time) that it could provoke distinct unease, if not physical illness, on confronting a breakfast sausage. Even though *Jungle* was to be his most famous book from more than 80 others and even though he performed a valuable service to the

American public, Sinclair remained bitterly disappointed in the public's reaction to the novel throughout the rest of his life. He once remarked ruefully that he had "aimed at the nation's heart, but hit it in the breadbasket."

In 1938 the Pure Food and Drug Law was comprehensively revised as the Federal Food, Drug, and Cosmetic Act (FFDCA), which is still today the cornerstone piece of legislation. The law created the Food and Drug Administration (FDA) as the regulatory body. The new law preserved the label as the regulatory device, but it was broadly redefined to include all descriptive materials that accompanied the product, such as packaging, inserts and advertisements. It prohibited the addition of poisonous substances to food, but when such additions were unavoidable or required during production, it empowered the FDA to set safe limits or tolerances for them. Such substances included the pesticides of the day, which were often various forms of arsenic, copper, or lead. Another example of an unavoidable contaminant originated with the solder used to join the seams in tin cans. Although the inside of the can is coated with shellac, some lead inevitably leaches from the solder through the shellac and into the contents of the can. At the time that FFDCA was passed there had been a proliferation of quack medical devices on the American market, and the law had tough new provisions aimed at these. The food part of the law provided the authority for factory inspections, and set up food standards to promote honesty and fair dealing in the interests of consumers.

The revision of the old Pure Food and Drug Law was prompted in part by the Elixir of Sulfanilamide scandal of the mid 1930's. Sulfanilamide was the first truly effective, systemic antibiotic to be discovered. Prior to that time the available drugs were either largely ineffective or too dangerous to be taken internally. There was every reason to believe that sulfanilamide would be a lifesaving drug in certain kinds of bacterial infections, and that there would be a large and lucrative market for it. Both for altruistic and economic motives there was considerable pressure to get the drug on the market. The drug itself had successfully passed the toxicity testing required at the time. A liquid formulation was thought to be desirable, but the drug was poorly soluble in water. Eventually a vehicle or solvent was settled upon that contained high concentrations of diethylene glycol. Unfortunately, diethylene glycol had not undergone toxicity testing.

Because diethylene glycol made no direct contribution to the antibacterial action of sulfanilamide, it could be listed on the label simply as an "inert ingredient." As in the case of some insecticide sprays with kerosene bases and some mouthwashes that contain high concentrations of alcohol, diethylene glycol turned out to be more toxic than the "active ingredient." It produced dangerous kidney damage, and in the form of the elixir, it was ultimately responsible for at least 100 human deaths in acute renal failure among patients treated with it. It is tribute to the USPHS that the instrument for this tragedy was unmasked as rapidly as it was.

The irony, and the circumstance that made pinpointing the blame so difficult, was that the elixir was especially valuable in and, most often prescribed for, infections of the urinary tract. Thus, the product was capable of producing the same end result, namely, renal failure, as the disease for which it was prescribed. As the iatrogenic illness superceded the bacterial infection, physicians assumed that treatment was inadequate, which led them to prescribe larger doses in the hope of gaining control. Instead, they were making things worse. Moreover, the elixir, as used therapeutically in regularly divided doses each day, maximized the already high potential of the solvent to produce cumulative effects. Finally, sulfanilamide itself was recognized to have a certain potential for producing renal damage. Sorting through these false leads and assumptions was truly a challenge to medical scientists.

In 1954, the Miller Pesticide Amendment was added to FFDCA to streamline procedures for setting safe limits for pesticide residues on raw agricultural commodities. The law empowered the FDA to establish maximum permissible levels for pesticides and other chemicals in food.

In 1958, the Food Additives Amendment was enacted to prohibit the use of any new food additive until the sponsor had established its safety. The FDA was then authorized to issue regulations specifying the conditions for their safe use. Food additives that were already in use at the time, some of them for many years, were "grandfathered" by placing them on a list that designated them as "generally recognized as safe" (GRAS list).

The current state of affairs is that the EPA is responsible for regulating pesticides and other environmental chemicals in food, whereas the FDA is responsible for food additives whether deliberate or introduced by necessity even with good manufacturing practice. The two agencies are empowered to establish either tolerances or action levels for the maximum permissible concentrations of these chemicals in foods. A tolerance is a regulation adopted through formal rule making procedures, requiring public hearings. A tolerance has the force of law and cannot be questioned by any court. An action level is a more informal judgement by the regulatory agency about the concentration of a given contaminant that will not adversely affect human health. It represents only a commitment by an agency to initiate regulatory enforcement action against those responsible for foods containing excessive levels, and it does not require public hearings.

A short list of tolerances and action levels as they existed in 1978 is shown in Table 8–1 for various foods. All of the chemicals

TABLE 8–1. EXAMPLES OF TOLERANCES (T) AND ACTION LEVELS (AL) FOR SOME INSECTICIDES IN SELECTED FOOD ITEMS.

Insecticide	Food Items (ppm)		
	Eggs	Dairy Products	Milk
Heptachlor/heptachlor expoide	0.03 (AL)	0.3 (AL)	0 (T)
Aldrin/dieldrin	0.03 (AL)	0.05 (AL)	0.3 (AL)
DDT/TDE/DDE	1.5 (AL)	1.25 (AL)	0.05 (T)
Benzene hexachloride	0.5 (AL)	0.3 (AL)	0.3 (AL)
Endrin	0.03 (AL)	0.3 (AL)	0.3 (AL)
Malathion	0.1 (T)	—	0.5 (T)

(From Babich H, Davis DL: Food tolerances and action levels: do they adequately protect children? BioScience 31:429, 1981. Copyright 1981 by the American Institute of Biological Sciences.)

in that list are insecticides of the DDT-type except for malathion, which is an organophosphate-type insecticide. Note that in some cases the tolerance was set not only for the parent insecticide, but for known metabolites as well, e.g., DDT/TDE/DDE. These values had originally been set by the FDA, but the authority for their enforcement was transferred to the EPA with the passage of FEPCA. The issue is more complicated, however, because different tolerances have been set for raw and processed foods. The EPA can use its own judgement to permit pesticide residues on raw food even if the chemicals have been shown to be carcinogenic, but if the residues concentrate during processing of the food, the new levels must be shown to be safe or the use must be banned. There seems to be no scientific rationale for this aspect of the law.

The Color Additives Amendment to FEPCA was added in 1960. This law allowed the FDA to regulate the conditions for the safe use of color additives or dyes in foods, drugs and cosmetics, and required the manufacturer to perform the necessary testing to establish their safety. Of all the various types of food additives, food dyes are among the most difficult to defend because they are used purely for aesthetic purposes. It is true that some foods would be less appealing in their natural colors, but food dyes have had a checkered record of safety. Several have been banned when they were shown to be carcinogenic.

Section 706B of the Color Additives Amendment was introduced by Congressman James J. Delaney, a Democrat from New York's seventh district. It was to become the famous "Delaney clause." In essence the Delaney clause banned from use in foods, drugs, and cosmetics any chemical that had been found to induce cancer when ingested by man or animals. Another way of stating it is that it sets an automatic tolerance of zero for any carcinogen in food. The assumption was that there is no safe dose (no threshold or NOEL) for carcinogens, an assumption that is still hotly debated by experts in the field. Because a negative vote on the Delaney clause could be viewed as evidence of a casual disregard for public safety, it passed by a large majority. It became law and it remains so today despite many attempts to strike it down.

The concept of a zero tolerance was embraced by the Congress, but it became a nightmare for scientists. From a philosophical standpoint one can see nothing but good in it, but many operational problems were rapidly recognized. The law did not specify the tests by which a carcinogenic potential was supposed to be identified, or the conditions (particularly the range of dosing levels) under which the chemicals were to be tested. In testing for carcinogenic effects, it is common practice to use doses that are much larger than those that might be encountered by ordinary environmental exposure. This practice often reduces the length of time needed to observe positive results and lessens the chances for obtaining false negative results. For those who believe, however, that there are safe levels of exposure even to carcinogens, it increases the chances for false positive findings.

The search for carcinogens in food at that time took on some of the aspects of a witch hunt. If one happened to be a scientist in the field seeking national recognition, an almost certain way to achieve that goal was to show that some widely used chemical found in food was carcinogenic. If a chemical is totally devoid of carcinogenic effects, no test procedure will yield positive results, but there are degrees of potency among cancer-causing chemicals just as there are for other toxic effects. In the cases of low potency carcinogens, positive results can sometimes be obtained if the testing is pushed to extreme levels of dosing. During this period it is probable that at least some chemicals were assigned zero tolerances when the particular test, the range of doses used, or the diagnosis based on microscopic examinations of tissues were questionable. On the other hand it is probable that some lives were spared by the banning of other chemicals.

Perhaps the most difficult problem involved the definition of a zero tolerance in a practical or operational sense. To the Congress it meant simply that no food should contain any carcinogen. But, it was left up to the analytical chemists to prove that none was present, and they were painfully aware that they had been assigned an impossible task. Any analytical procedure for any chemical has a finite and defined detection limit. Below that limit the chemical could still be present in food, but it would be "invisible" to that particular analytical technique. The chemists would have no choice except to say that as far as they could see the zero tolerance was being met. It would not, however, be the zero tolerance as intended by the Congress.

Although DDT is not a carcinogen, it will be used as a convenient example to illustrate this operational problem with the Delaney clause with the understanding that the same principles could apply to an actual carcinogen. One modern, sensitive, analytical technique has detection limits of 1×10^{-9} gram (1 nanogram) of DDT in a 100-gram sample of milk. At the same time, however, DDT could be present in the amount of 2×10^{12} molecules in the same sample and escape detection. In the view of the "no safe threshold" school of thought even a single molecule of a carcinogen is believed to be able to initiate the disease. The detection limits of the above method for DDT permit a huge number of molecules to escape detection. If the no safe threshold concept is valid, a zero tolerance fails to protect the public, and instills a false sense of security.

Figure 8–1 illustrates the detectability limits for DDT by various analytical techniques on a log scale that spans six orders of magnitude (1 million-fold) over the years since DDT measurements have been made. At the time that the Delaney clause was enacted, DDT was measured by a colorimetric technique that was sensitive only to 25 μg (25×10^{-6} g). Although that was regarded as state-of-the-art at the time, no one could have predicted the revolutionary advances in analytical technology to come, which would bring previously undreamed of sensitivity into otherwise quite ordinary laboratories. The least detectable amount of DDT became steadily smaller over the years through the development of gas-liquid chromatography (GLC), paper chromatography, thin layer chromatography, polarography, and new GLC detector systems. By historic accident the ultimate in sensitivity, the electron capture detector, which measures DDT in the picogram range (1×10^{-12} g), was discovered relatively early. It improved the sensitivity of detection by 250 millionfold. Inevitably, however, the possibility that DDT might still be present at some concentration below the detection limit, no matter how low that is, always exists.

This dilemma, as pointed out by Zweig, is that now one can bring to bear exquisite sensitivity of detection to pollutants that may be ubiquitously distributed on the planet. If DDT had been carcinogenic, during the 1960s it would have been impossible with the new technology to find a food that was not contaminated. If DDT had been assigned a zero tolerance under the Delaney clause and the clause had been strictly enforced, mankind would have been doomed to starvation. From that perspective, the Delaney clause begins to look somewhat ridiculous, but it is still the law.

In recent years it has been suggested that the Delaney clause is outmoded because its inflexibility does not permit consider-

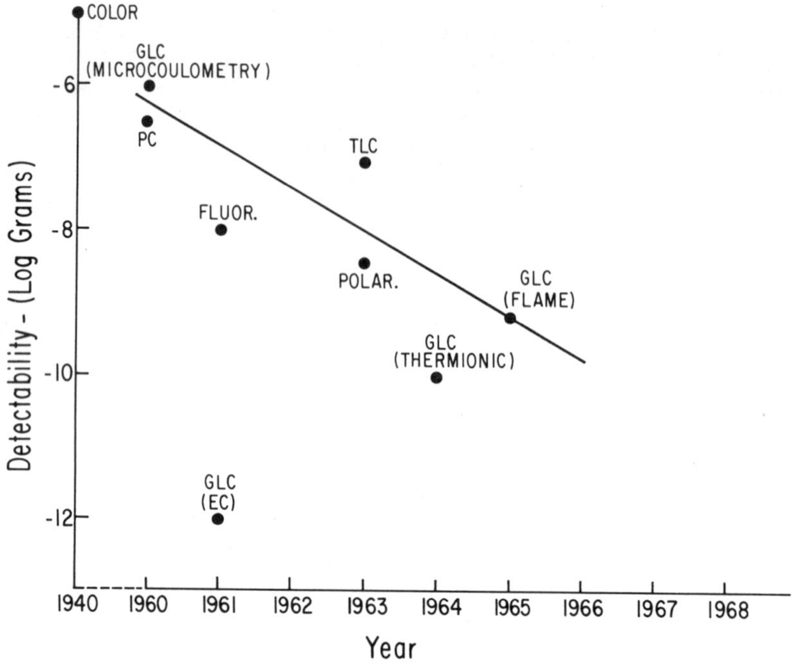

FIGURE 8–1. Technologic advances by year of their introduction, which resulted in increased sensitivity for the measurement of DDT. (See text for additional details.) (After Zweig G: The vanishing zero: the evolution of pesticides analysis. *In* Blood FR (ed): Essays in Toxicology, Vol. 2. New York, Academic Press p. 155, 1970.)

ation of the issue of relative risk. Central to this issue is the question of what the public regards as an acceptable risk. Would the public be willing to regard a risk of one in one million as insignificant? The chemicals that have stirred the debate are a half-dozen food dyes, five reds and one orange, that are proven animal carcinogens. Industry representatives feel that the risks of the most widely used dye, Red No. 3, are on the order of one in 31 million, but the Delaney clause would clearly ban it. Decision after decision has been delayed by legal maneuvering and requests for further study. At present, the only alternative, which led to the exception of saccharin from Delaney, is a congressional mandate.

A final legal tangle emerged when dual standards were set for pesticides registered before 1978 and those that came up for registration thereafter. A strict interpretation of the Delaney Clause and a more extensive battery of tests were imposed on products registered after 1978. A paradox emerged in which a new pesticide with a risk of 1 in 10 million to 1 in 100 million of causing cancer would be denied registration when it could replace an older pesticide with a risk of 1 in 10,000 to 1 in 100,000. Pesticides are well known to be present in ground water and in potable water supplies, yet this source of exposure is not taken into account in regulating pesticide use. Indeed, the EPA has proposed transferring the regulation of pesticides in ground water to the state level. Clearly, these situations make no sense even to the lay public, and sweeping reforms are needed.

It is the farmer or food producer who must operate under the threat that residues of pesticides on his crop after harvest might exceed the EPA tolerances. If such is found to be the case, his crop may be confiscated at the time that it crosses a state line. Many states, however, have their own laws for crops moving in intrastate commerce which were patterned after the federal legislation. Unfortunately, the farmer has little control over a number of critical events. Pesticides are costly because of the research and testing required for their registration. If the farmer tries to cut costs by skimping on their application, his crop may be ruined by insects or uncontrolled weed growth, but if he uses too much, his crop may be confiscated because it exceeds the tolerance. All the farmer has to go on is a set of EPA-approved directions for applying a particular pesticide to a particular crop. Many uncontrollable factors, such as rainfall, temperature, wind, or sunlight, may influence the final concentration of pesticide residue on the crop.

The set of directions can only be derived by actual crop experience, and both the pesticide industry and the federal establishment maintain elaborate experimental farms in order to acquire that experience. These add additional costs to the development of the final product. Laboratories of the EPA are also constantly engaged in analyzing samples from food crops moving in interstate commerce to ensure that the tolerances are not being exceeded. Actual analyses are conducted on what authorities consider to be an adequate sampling of all food commodities.

In order to generate data on actual dietary intake of food contaminants one needs to know what kinds of foods Americans eat and how much of each kind. Since 1965 the FDA has been monitoring selected foods throughout the United States in its Market Basket Survey Program. A parallel effort was begun in 1965 by the Department of Agriculture. These surveys included only 15- to 20-year old American males, a group that consumes more calories than any other with the possible exception of professional or olympic athletes. They would, however, also consume the greatest amounts of any contaminants present. Over a two week period the diet of this group included 117 distinct food items. Periodically, these market basket items are also purchased locally and analyzed for contaminants by composite food classes, such as dairy products, leafy vegetables and root vegetables.

Table 8–2 contains a glossary of terms that will be useful as we examine how toxicity data, actual food intakes, and tolerances are used to make judgements about safety. First, critical data must be generated from a 90-day feeding study to establish a NOEL. These studies are done in a minimum of two species. The most common criterion for an adverse effect (toxic level of feeding) is failure to gain weight appropriately, but sometimes changes in organ weights, e.g., liver weight to body weight ratio, or effects on enzyme activity, e.g., cholinesterase, may be used depending on what is known about the pesticide and its anticipated toxic effects. Next, the NOEL for the most sensitive species of the two tested is divided by an arbitrary safety factor, which often has the value of 100. This factor is based on the assumption that animal-to-animal variation within the most sensitive species may be as large as tenfold, and that humans will be 10 times more sensitive than the most sensitive animal species tested. For cholinesterase inhibitor type insecticides, such as malathion, considerable experience has shown that humans and most laboratory animal species are about equally sensitive so that a safety factor of only 10 is adequate to protect the public. In setting a temporary tolerance when the animal data is incomplete, a safety factor of 2000 may be used.

Assume that the manufacturer of a new herbicide wants to register his product with the EPA. The 90-day studies in rats and mice have shown that rats are the more sensitive species. The NOEL in the rat study was 300 ppm (300 mg/kg of diet). The average rat in the study consumed 20 g of diet per day and weighed

TABLE 8–2. GLOSSARY OF FOOD SAFETY TERMS, ACRONYMS, AND STANDARD VALUES

ADI (acceptable daily intake) = NOEL/safety factor from 2 to 2000
Average body weight of consumer = 60 kg for EPA and 70 kg for FDA
Average weight of food consumed daily = 1.5 kg
Food factor = percentage of a standard diet made up by the food in question
Food tolerances = MPI/(food factor × 1.5 kg)
MPI (maximum permissible intake) = ADI × average body weight of consumer
NOEL (no observable effect level) = level of feeding (or dosing) that evokes no adverse response in the most sensitive test species
Safety factor = usually 100, but 2000 for temporary tolerances
TMRC (theoretical maximum residue concentration) = the sum of all the food tolerances multiplied by all their respective food factors

(From Babich H, Davis DL: Food tolerances and action levels: do they adequately protect children? BioScience 31:429, 1981. Copyright 1981 by the American Institute of Biological Sciences.)

200 g (0.2 kg). The NOEL can be translated into actual daily intake by:

$$\text{NOEL} = 300 \text{ mg residue}/1000 \text{ g diet} \times 20 \text{ g diet/day} = 6 \text{ mg residue/day}$$

$$6 \text{ mg residue per day}/0.2 \text{ kg} = 30 \text{ mg residue/kg body weight per day}$$

This NOEL for rats is translated into an acceptable daily intake (ADI) for humans by dividing by a dimensionless safety factor (SF). If the manufacturer is applying for a temporary tolerance:

$$\text{ADI} = (30 \text{ mg residue/kg per day})/2{,}000 \text{ (SF)} = 0.015 \text{ mg residue/kg per day}$$

The maximal permitted intake (MPI) is obtained by multiplying the ADI by human body weight, which differs between the EPA (60 kg) and the FDA (70 kg):

$$\text{MPI} = 0.015 \text{ mg residue/kg per day} \times 60 \text{ kg} = 0.9 \text{ mg residue/day per adult}$$

The tolerances set by the EPA and derived by actual crop experience when the pesticide was applied in accordance with good agricultural practices were 0.1 ppm for corn and 0.02 ppm for carrots. Tables of so-called food factors derived from surveys show that corn makes up 1.19% of the average American diet and carrots make up 0.55% of the diet. The theoretical maximum residue concentration (TMRC) is the sum of all the products of the tolerances and the food factors for each crop on which the pesticide is approved for use:

$$\text{TMRC} = (0.1 \text{ mg/kg} \times 0.0119) + (0.02 \text{ mg/kg} \times 0.0055) = 0.0013 \text{ mg/kg diet}$$

The average American consumes 1.5 kg of food per day; therefore,

$$\text{TMRC} = 0.0013 \text{ mg/kg diet} \times 1.5 \text{ kg food} = 0.00195 \text{ mg/day per adult}$$

Because the MPI (0.9 mg/day per adult) exceeds the TMRC (0.00195 mg/day per adult) by more than 400-fold, the temporary tolerances are acceptable, and the application should be approved.

In addition to food additives, pesticide residues, and chemicals unavoidably added during manufacturing, tolerances or action levels also may be set for industrial chemicals that are accidental contaminants. For example, polychlorinated biphenyls (PCB) are not pesticides and they are never deliberately added to food, but they have been found in many food products because of extensive environmental contamination.

SUMMARY POINTS

The Federal Environmental Pesticide Control Act of 1972 grouped pesticides, air and water pollutants, solid waste, noise, and radiation under the regulation of the EPA.

The EPA adopted the stratagem of using the registration as the regulatory device; every pesticide product moving in interstate commerce must have a valid registration.

The Federal Food, Drug and Cosmetic Act of 1938 established the FDA as the regulatory agency for those products, and it used the label as the regulatory device.

The Food Additive Amendment of 1958 banned new additives until their safety had been established; all the additives used up to that time were placed on a GRAS list.

Section 706B of the Color Additives Amendment of 1960, known as the Delaney clause, banned the use in food of any chemical carcinogenic to man or animals.

The setting of safe tolerances for pesticide residues permitted in food is an area in which science and the law come together, not always to their mutual satisfaction.

BIBLIOGRAPHY

Babich H, Davis DL: Food tolerances and action levels: do they adequately protect children? BioScience 31:429, 1981.

Geiling EMK, Cannon PR: Pathological effects of Elixir of Sulfanilamide (diethylene glycol) poisoning, clinical and experimental correlations: final report. JAMA 111:919, 1938.

Sinclair U: The Jungle. New York, Airmont Publishing, 1965.

Young JH: The Toadstool Millionaires. Princeton, NJ, Princeton University Press, 1961.

Zweig G: The vanishing zero: the evolution of pesticide analyses. In Blood FR (ed): Essays in Toxicology, Vol. 2. New York, Academic Press, p. 155, 1970.

9

CHAPTER

SURVEY OF PESTICIDES

Pesticides are also called "economic poisons" because their toxic properties are exploited commercially. They are used in a variety of ways to increase yield, improve quality, or promote consumer acceptability of a wide variety of consumer products. The major classes of pesticides are insecticides, rodenticides (mouse or rat poisons), herbicides (to kill weeds among food or ornamental plants), and fungicides (to eradicate or prevent mold growth on food or nonfood products). Poisons and repellants have been marketed for use against various pest amphibian and reptilian species, birds, fish, and invertebrate animals, such as slugs and snails. Nematocides are used against certain worms (nematodes), and herbicides can be subdivided into plant regulators, defoliants, and desiccants. Fumigants are often highly toxic gases or vapors that may be used against either insects or rodents or both. Modern agriculture, with its high yield per acre of good quality food, depends heavily on the intensive use of pesticides and fertilizers. The American consumer not only expects, but demands, foods of high quality. For most of us, the days in which we could bite into an apple, find half a worm and cheerfully continue to eat the fruit are long past.

FUNGICIDES

Both humans and animals are susceptible to topical fungal infections; athlete's foot is a familiar example. Less commonly, dangerous systemic fungal infections may develop (mycoses), which are analogous to bacterial infections except that they are not transmissable from one individual to another. These may be acquired by inhaling fungal spores, or they may be caused by an overgrowth of normal fungi that are present in the body. Mycoses can be treated with topical or systemic antifungal drugs, but these are different chemicals from the so-called fungicides. The pesticide-type fungicides are used to prevent or control mold growth on feeds, seeds, leather, paper, fabrics, cosmetics, and plants. Their use is not simply for aesthetic purposes. Many of the above products can be ruined by mold or mildew, and economic losses in tropical climates from that cause are substantial.

Another use for fungicides is to prevent the growth of certain fungi that synthesize highly toxic or carcinogenic chemicals called mycotoxins. In some cases the mycotoxins are found within the mold so that the fungus itself must be consumed in order to produce poisoning. In other cases the mycotoxins are released into the food so that even if the mold is removed, poisoning may result. It has proven very difficult to try to define the magnitude of the human health problem, if any, caused by mycotoxins in developing tropical countries. Only fragmentary evidence exists to suggest that they may be a cause of human illness or death. Mycotoxins have certainly been responsible for epidemics of poisoning among commercially raised fish, fowl, and other livestock when they have contaminated the animal's feeds. If there were other alternatives, most humans would reject moldy food, but few would have qualms about feeding it to livestock.

Organomercurials, such as methyl mercuric chloride (CH_3-Hg-Cl), are fungicides often used on food crop seeds to protect them against molds that might impair their viability. In the fall and winter of 1971–1972 an epidemic of poisoning caused by methyl mercury occurred in Iraq, in which 6530 persons were hospitalized and almost 500 died. The source of the exposure was bread prepared from flour made from wheat grain intended for use as seeds only. There would have been no danger in eating wheat grown from methyl mercury treated-seeds because the subsequent grain crop would contain little or none of the fungicide. The Iraqi peasants, however, were starving, and even though the seeds had been dyed red to serve as a warning, they were promptly milled into flour and consumed as bread.

Methyl mercury produces a devastating human illness affecting primarily the brain with many neurologic disturbances including paralysis, "tunnel vision," and blindness. There is no effective antidote as there is for inorganic mercury salts, nor are there any truly efficacious means for hastening its excretion from the body. Unfortunately, methyl mercury is also highly teratogenic. Offspring exposed *in utero,* if they survive, may have an irreversible affliction resembling cerebral palsy. Experimentally, methyl mercury has been shown to be mutagenic as well.

Two outbreaks also occurred in Japan, in Niigata and in the area surrounding Minamata Bay. These are of particular interest because the source of the original pollution was not methyl mercury itself, but inorganic mercury wastes discharged into the bay by surrounding industries. Anaerobic bacteria, which are able to live without oxygen in the detritus on the ocean floor, were able to convert this inorganic waste into methyl mercury. The methyl mercury was concentrated as it was passed upward through natural food chains, and it found its way into fish and shellfish, which were consumed by families living on the shores of the bay.

The population of the United States has been fortunate in avoiding mass poisonings in the past, but there was one isolated series of cases involving a single family in Almagordo, NM. The father worked in a seed store that supplied local farmers, and he maintained a few pigs at home. He noticed a significant amount of wastage in the form of spilled seed grain at the store, and he began sweeping it up to feed to his pigs. Within a short time his pigs became obviously ill. Fearful of the loss of his investment, the father had them butchered, and he froze the meat for the use of his family. Three of them were eventually poisoned severely.

Between 1955 and 1959 southeast Turkey suffered an outbreak of poisoning, again involving fungicide-treated grain. In this case the fungicide was hexachlorobenzene (HCB), a chemical that has many properties in common with the halogenated hydrocarbon insecticides (see Chapter 10). Almost 3000 people were poisoned, 300 of them fatally. These victims of HCB poisoning resembled closely victims of a rare inherited disease called porphyria cutanae tarda. Their skin became unusually dark (hyperpigmentation), they had an overgrowth of body hair, there were signs of liver damage, and some became frankly psychotic. The bizarre appearance of some young patients gave rise to the name, "monkey children." HCB is one of a number of chemicals that are secreted in substantial amounts in breast milk, and nursing infants were poisoned by that route instead of by consuming the contaminated grain.

Pentachlorophenol ("penta") is a fungicide that is also marketed as a molluscicide, germicide, and wood preservative. Molluscicides are used in developing tropical countries to control snails, which are carriers of the parasite that causes schistosomiasis. Penta as a wood preservative is one of the chemicals of concern in indoor air pollution. During the period between April and August, 1967, twenty infants in a midwestern hospital nursery developed an unusual illness characterized by sweating, high fever, rapid heart rate, and acidosis. Nine were severely ill and two died.

After an exhaustive search for the source and identity of the suspected toxic chemical, it was traced to a product found in the hospital laundry. The product was clearly labelled for use on hard surfaces only (e.g., floors, sinks, counter tops,). The label also contained a specific warning against its use as a laundry detergent. Despite these danger flags, it had been used to launder infant clothing and bed linens. When these items came in contact with infant skin, which is unusually permeable to some chemicals, the penta in the product was absorbed systemically to produce poisoning. Penta was mentioned earlier (see Chapter 3) as an example of a chemical that is more hazardous when the ambient temperature is high because, as noted above, one of its toxic signs is extreme fever.

Dozens of other types of chemicals are marketed in the United States as fungicides, and some of these are more widely used than the three examples above. These were selected because they had been the cause of multiple human poisonings, involved unusual sources of exposure, or illustrate how human error contributed to tragedy. These three fungicides represent the extremes in toxicity; most fungicides are reasonably safe when used in accordance with the manufacturer's recommendations.

RODENTICIDES

Rodenticides are used in attempts to control populations of rats, mice, and other rodents that destroy enormous quantities of food and spread dangerous human communicable diseases. A truly effective worldwide program of rat control would have a significant impact on food supplies in developing nations. Fully 20% of the world's food crops are lost to rats each year. Unfortunately, the rat is a remarkably resilient creature. It survived atomic bomb testing on Engebi and other Pacific islands of the Eniwetok Atoll. They reproduce so rapidly that a single mating pair could have 15,000 descendants in a year. Second only to humans, rats are the most successful and numerous mammals on earth. The parasitic insects that live on them are responsible for the spread of a score of human diseases including louse-borne typhus and the bubonic plague ("black death"), which was responsible for the demise of one-quarter of the population of medieval Europe. In at least one way, however, rats have partly repaid their debts to society. They are the most widely used species in biomedical and toxicologic research.

In years past, violently toxic chemicals were marketed and used as rodenticides, including cyanide, arsenic, phosphorus, strychnine, and thallium. Most, if not all of these, are now banned in American products. Rat poisons are particularly effective because of a species peculiarity of rats: they cannot vomit. Vomiting, when it occurs in humans, may be self-protective because it limits the absorption of poisons taken by mouth. Plaster of Paris was an early rodenticide; when consumed in adequate amounts, it will "set" in the gastrointestinal tract resulting eventually in a fatal obstruction. Manufacturers are constantly searching for chemicals that are more or less selectively toxic to rodents.

During the winter of 1921–1922 cattle herds in the upper midwestern states and the adjoining provinces of Canada were decimated by a new disease. The course was rapid and its termination abrupt. Mortality rates in some herds ran as high as 90%. The characteristic feature was uncontrolled hemorrhaging, sometimes externally, but more often internally. An infectious disease was quickly ruled out as the cause. As had been the practice for many years, all of the cattle had been fed on sweet clover hay. It had, however, been a particularly wet harvest season, and many farmers had experienced difficulty in adequately drying their hay crops. Much of it was visibly moldy. Indeed, only moldy hay produced the disease, but not all moldy hay did so. Suspicion began to center around particular kinds of molds and perhaps some product found in them or synthesized by them.

Ten years went by before it was discovered that the bleeding tendency in cattle was caused by a deficiency in prothrombin. Prothrombin is a protein synthesized by the liver and released into blood plasma. It is one of a number of factors responsible for normal blood clotting. If supplies of prothrombin are deficient, any minute injury to blood vessels whether external or internal becomes a potential site of hemorrhage because the body is unable to plug these with tiny clots. Eventually it was found that certain molds growing on hay were able to convert a naturally occurring chemical to a potent anticoagulant substance, dicoumarol. Dicoumarol is an anticoagulant because it interferes with the ability of the liver to make prothrombin.

A number of synthetic derivatives of dicoumarol were then prepared. One of these was about 50 times more potent than dicoumarol as an anticoagulant, and it was effective in the body in about half the time that it took for the effects of dicoumarol to appear. It was named warfarin because the research was funded by the Wisconsin Alumni Research Foundation. Warfarin rapidly became a useful drug to prevent excessive and dangerous clot formation in patients at risk for strokes or at risk for clots that might lodge in the lungs (pulmonary embolism). At almost the same time it was recognized as a useful rodenticide, and it is now the active ingredient most widely used in such products including the D-Con line. As already noted (see Chapter 4) warfarin has a moderate tendency to produce chronic poisoning. It acts slowly because the body's stores of prothrombin must be depleted before the deficiency state can occur.

The resilience of rats as a species, however, quickly became a factor of some importance. Warfarin-resistant strains of rats were recognized as early as 1958, and this resistance could have seriously limited its effectiveness. Scientists, however, have already developed a chemical that potentiates the anticoagulant effect of warfarin. It is a chloro-derivative of vitamin K. Ordinarily, vitamin K is an antagonist to warfarin poisoning because it stimulates the liver to synthesize prothrombin. The chloro-derivative of vitamin K prevents the antagonistic effects of genuine, dietary vitamin K, and when given in combination with the rodenticide, it results in a superwarfarin. Unquestionably, the search will continue for new rodenticides. One such chemical, norbormide, is marketed under the trade names, Raticate and Shoxin. These brand names are apropos because its lethal effect in rats is related to an intense, extreme, and irreversible vasoconstriction.

HERBICIDES

The earliest herbicide may have been ordinary table salt (sodium chloride). When spooned tediously into cracks in brick sidewalks for example, it was effective in reducing the infiltration of crab grass. In high concentrations it kills plants by drawing water out of them (a dessicant or drying agent). Almost all modern systemic herbicides act by one or the other of two fundamental mechanisms: (1) they suppress plant growth or they stimulate growth in an uncontrolled and self-destructive manner, or (2) they interfere with photosynthesis. Because photosynthesis occurs only in green plants, it might seem that herbicides of the second type could be devised that were highly selective for plants and relatively nontoxic to animals. Unfortunately, such has not proven to be the case.

Systemic herbicides are absorbed and translocated within plants. The development of chemicals that influence plant growth had to be preceded by the discovery that plants, like animals, have hormones that regulate the rate of their growth. The fundamental observations on this point were reported by Charles Darwin in the 1880s. He noted that plants exhibited primitive growth patterns called "tropisms" in response to certain stimuli. Phototropism refers to the well known tendency of plants to grow toward a source of light when the ambient illumination is uneven. The existence of geotropism, the tendency for plants to orient themselves upright in a gravitational field, was long suspected and finally proven through the technology of space exploration. When plants are grown in zero gravity centrifuges or space satellites, they no longer know which end is up.

Early botanists believed that some chemical substance found in plants was responsible for controlling the direction and perhaps the rate of growth. Important discoveries in this area were made with the lowly bean sprout. The shoot or sprout has neither buds nor leaves. It is called a coleoptile, and its simple structure has made it a favorite subject for experiments on plant physiology. Beans only sprout in the dark, and if they are moist. In a dark room their behavior can be observed under an ordinary photographic safe light with controlled conditions of temperature and humidity. By making ink marks at regular intervals along the length of the coleoptile, one can after time, see which part of the sprout is growing most rapidly. Surprisingly, the tip grew very little. The region of most rapid growth occurred about a third of the way down from the tip.

Even though the tip did not grow, if one slices off as little as 2 mm of it, the growth of the coleoptile stops. If the tip is placed back onto its stump while it is still viable, growth resumes with virtually no change in rate. A Dutch botanist named Fritz Went in the 1930s, became convinced that a chemical substance must be manufactured in the tip and secreted onto cells below that controlled the growth of the coleoptile, and he proved his point with an experiment that was breathtaking in its simplicity.

As illustrated in Figure 9–1, Went cut off the tip of a coleoptile and placed it on a block of agar. Several hours later he discarded the tip and placed the block of agar on the stump, but he placed the agar block off center. Growth of the stump resumed, but only on the side of the sprout where the agar block had been placed so that the coleoptile grew in a curved fashion. The plant cells must have grown faster on the side with the agar block. The curvature of growth was analogous to the turning of a military tank when the treads on one side turn faster than those on the other. Clearly, a growth factor had been transferred from the tip to the agar block and then from the agar block to the stump. This one simple experiment not only proved the existence of plant growth hormones (now called auxins), but it showed also that auxins account for tropisms as well. Much still remains to be learned about auxins, however, because they actually appear to inhibit growth in roots, and calcium seems to play an important but as yet undefined role in their actions. Naturally occurring substances have recently been found in plants that regulate the transport of auxins. These could importantly influence the sites where auxins exert their effects within the plant.

Eventually, the first auxin was isolated, purified, and its chemical structure determined. The stimulus was at hand for chemists to make hundreds of synthetic derivatives. One day it was noted that when one such derivative had been sprayed on a grassy plot, it killed all the broadleaf plants that were also growing nearby without affecting the growth of the grass. Thus, the concept of a selective herbicide was born, i.e., a chemical that selectively de-

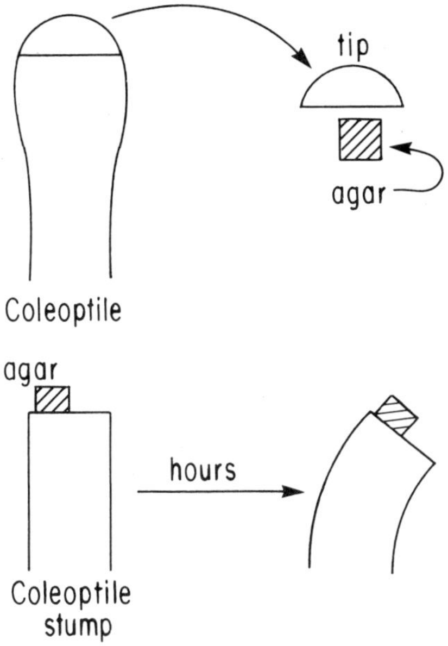

FIGURE 9–1. The experimental approach that proved the existence of plant growth hormones (auxins) and showed how they are responsible for tropisms. (See text for details.)

stroys broadleafed weeds but spares grasses or cereals, which constitute the major food crops.

The first large scale demonstration of the potential for selective, systemic herbicides was on a sugar cane field in Puerto Rico in 1944. The chemical was known as 2,4-D. Later, a close chemical relative, 2,4,5-T, was to be discovered that was especially good for defoliating woody plants. The results, as Samuel Johnson might have put it, "exceeded the wildest dreams of avarice." It was effective at low concentrations, and at a cost at that time of only about a dollar an acre. It eliminated the need for tedious hand or even mechanical weeding. It was relatively nontoxic to animals, but even today the exact nature of its toxic effects in mammals is poorly understood.

Through the early 1950s the market for 2,4-D increased more rapidly than that for any other pesticide. Several hundred brands and closely related chemicals are on the American market, and new uses have been discovered. Instead of spraying it on growing weeds, it was found that if it was applied to soil in higher concentrations, it could prevent weed growth from starting (preemergent herbicide). A mixture of equal parts of derivatives of 2,4-D and 2,4,5-T were later to become the active ingredients in

Agent Orange (see Chapter 11). In plants 2,4-D appears to act like a super auxin to cause plant death through uncontrolled and exhaustive growth. Later herbicides were to be developed that act by blocking the effects of natural auxins to suppress growth.

The triazine herbicides are a family of chemicals that includes atrazine, and they act by inhibiting photosynthesis. Photosynthesis is the chemical process by which plants use solar light energy, carbon dioxide (CO_2), and water to make essential nutrients and eventually structural components of their cells. Atrazine provoked the first massive seizure of food by the FDA under the provisions of the Delaney clause (see Chapter 8), which prohibited the inclusion in foods of any chemical that had ever been shown to cause cancer in animals or humans. Atrazine did produce tumors when fed in high doses to rats. It had been widely used on cranberry crops in New Jersey, and trace residues had been identified in some cranberry products. One week before Thanksgiving in 1959, the FDA moved to seize all of that year's cranberry crop most of which was already in the form of sauce. Events in the food industry removed from that point in time can be dated as BC or AC, before or after cranberries. The seizure was a crippling financial blow to the cranberry industry, and the first test of the power of the FDA to evoke the Delaney clause. Since then it has been used in as few as a half-dozen seizures, but its threat is a constant presence in the food industry.

Again, there are dozens of other chemical classes of herbicides, but the one with the least enviable record of safety is paraquat, a member of the bipyridilium family. These are also inhibitors of photosynthesis. Paraquat can be highly acutely toxic to animals and humans, but it does not have a significant tendency to produce chronic poisoning. More than 200 human deaths have resulted from the accidental or deliberate ingestion of aqueous paraquat concentrates (20 to 40% active ingredient). It is far less hazardous when diluted to a concentration appropriate for its use as a herbicide.

Irrespective of the route of administration, the terminal events in paraquat poisoning involve the lung. Multiple organs are damaged, however, including the liver, kidneys, and heart although these may show signs of recovery even before the lethal damage to the lungs begins to appear. Victims may survive for a month before finally succumbing. Paraquat destroys the fine structure of the lungs, and this is replaced by fibrous connective tissue (fibrosis) that interferes with the absorption of oxygen from air. The end result is a slow, relentless suffocation. No effective antidotes are known, and poisoned patients are difficult to treat effectively. The illness is so desperate that lung transplants have been attempted, so far unsuccessfully. As many as one-half of the reported cases have ended in death. Ironically, the one thing that the body needs for survival, namely, oxygen, actually intensifies the lung damage. The basis for both paraquat's herbicidal activity and its toxicity to humans is its ability to change molecular oxygen into much more reactive and toxic forms called free radicals. Experimentally, it has been shown that survival is prolonged if the oxygen content of the air breathed by poisoned animals is lowered instead of instinctively raised.

For a brief period of time paraquat was sprayed from airplanes onto illicit marihuana in Mexico that subsequently found its way into the United States. This practice has since been halted. No human intoxications resulting from paraquat have ever been reported from that particular source of exposure although traces of the herbicide have been identified in marihuana found in this country. There are dangers to marihuana, but paraquat may have been the least of these.

INSECT REPELLANTS AND AVICIDES

Folk nostrums used to repell insects include oil of citronella for mosquitos and Vicks Vaporub for black flies. Their pungent odor, however, makes them socially unacceptable for what may be only marginal efficacy. Avon's Skin-So-Soft smells good, but does little against biting insects. Two modern insect repellants are Rutgers 6-12 and Off. These are odorless to humans, and there is some sound evidence that they actually work. They were discovered, however, during large scale screenings of chemicals of all types for that particular property, and how they work on a molecular basis is unknown. A few human poisonings have resulted from applications of Off. It is suspected that these occurred in genetically predisposed persons.

Avicides, or bird poisons, are among the more unusual types of pesticides, but localized pest infestations of starlings, sparrows, or crows can constitute a significant nuisance and result in substantial economic losses of food crops. At least one product, Starlicide, is highly selective for some but not all species of birds. It appears to produce kidney damage in birds but not mammals. Such products are unlikely to become popular because they are almost certain to provoke an environmental backlash. It is one thing to wage war on mice and mosquitos, and another entirely to combat even pest birds.

INSECTICIDES

The insecticides are the most important pesticides from a commercial standpoint. The insecticides prior to World War II were of plant origin, and they fell into two categories: (1) those that were effective but highly toxic—concentrated solutions of nicotine were widely used at one time; few poisons are more rapidly acting, and it easily penetrates the intact human skin, and (2) a second group that were much safer but far less effective. These include pyrethrum, which is the dried powdered flowers from a plant of the *Chrysanthemum* family. It contains at least six active ingredients which are known as pyrethrins, and many synthetic and semisynthetic derivatives have been prepared, which are called pyrethroids.

Another natural insecticide in the second category above is rotenone, which is derived from a plant known as cube, native to Central and South America. The plant contains other active chemicals, but rotenone is the most potent. Again, many synthetic derivatives have been prepared. Rotenone is highly toxic to fish and birds, and it has been used by natives as an arrow poison. As so-called "fish berries" it may be distributed in open bodies of water to acquire fish for food or to eradicate unwanted species prior to restocking lakes or ponds with game fish.

Both pyrethrum and rotenone are still widely used today. They are among the most popular ingredients in household insect sprays. Obviously, they are considered safe in normal use, and, indeed, they are relatively nontoxic when given by mouth to animals. Both are highly toxic, however, if injected. The pyrethrins are rapidly inactivated both by insects and by mammals. Rotenone is only slowly and incompletely absorbed when given by mouth to animals, and that fraction which is absorbed is metabolized both to active and to inactive products.

Interest was renewed in both pyrethrum and rotenone when a way was discovered to make them more effective as insecticides. Modern products contain not only the active ingredients, but so-called synergists as well, such as piperonyl butoxide. The active ingredients undergo biotransformation by P450-like enzymes in target insects to inactive products. The synergists act to prevent the metabolic breakdown of the active ingredients; therefore, they potentiate their toxicity. Synergists increase the toxicity of these insecticides to mammals as well, but they are still considered safe enough for proper home use.

There are a large number of so-called organophosphate insecticides (OPI) ranging from the supertoxic parathion to malathion, which has been approved for household use. These chemicals are related to the nerve gases developed during World War II. Despite a wide diversity of chemical structures, they all act in the same way: by producing a long-lasting and sometimes irreversible inhibition of the enzyme, acetylcholinesterase. Acetylcholinesterase (AChE) is found at all nerve-to-nerve, nerve-to-skeletal muscle, and nerve-to-gland, smooth or heart muscle junctions where acetylcholine (ACh) is the neurotransmitter. When these nerves are stimulated, they release ACh, which crosses the synaptic cleft to activate various kinds of receptors on the other side to produce characteristic effects (see Chapter 1). AChE breaks down ACh to terminate the responses, but if AChE is inhibited by OPI, the actions of ACh are greatly and dangerously amplified.

The results of an uncontrolled flood of ACh differ from one receptor type to another. Glands are stimulated to secrete copious amounts of fluids. The secretory glands in the airways to the lungs may release so much fluid that gas exchange is impaired. The bronchi may be dangerously constricted. Oddly, however, instead of producing sustained skeletal muscle contraction, a flood of ACh on those receptors results in their desensitization and flaccid paralysis. The result is analogous to poisoning by curare (see Chapter 1), except that it results from excess ACh on the receptor instead of the "deficiency" caused by curare. In both cases the respiratory muscles are paralyzed. These and other effects of the OPI, including an action on the brain, all conspire to produce a failure in breathing.

Fortunately, effective antidotes are now available for the management of victims of nerve gas or OPI poisoning. The drug, atropine, will block the overproduction of secretions, and the effects on smooth muscle and heart. A chemical known as 2-PAM was specifically designed to reactivate inhibited AChE. Modern methods for the mechanical support of breathing and good nursing care all greatly improve the chances for survival.

A related and generally less dangerous group of insecticides are the increasingly popular carbamate type AChE inhibitors including carbaryl (Sevin) and aldicarb (Temik). These are referred to as reversible inhibitors of AChE, and their effects are generally less intense and of a shorter duration than the OPI. Otherwise, they act by the same mechanism. Unlike the DDT-type insecticides, none of the agents discussed here constitutes a significant problem because of environmental persistence, with the possible exception of their persistence in ground water and deep aquifers.

A poisonous volatile gas, methyl isocyanate, is used to make both Sevin and Temik. In December, 1984, a tank containing 45 tons of methyl isocyanate at a chemical plant in Bhopal, India, leaked the gas for about an hour, resulting in the worst industrial chemical disaster of all time. More than 2500 people died, and 150,000 victims were treated at nearby hospitals and clinics. Most of the deaths resulted from the lung irritant properties of the gas, which caused pulmonary edema. A number of the survivors were permanently blinded or sustained severe injury to nasal and bronchial passages. Some of the signs in poisoned victims suggested the presence of another chemical such as phosgene, a gas used in World War I as a pulmonary irritant, which is currently used to make methyl isocyanate.

Aldicarb is more acutely toxic than carbaryl, making it one of the most poisonous of all pesticides, but it is virtually devoid of chronic effects. It was one of the first important systemic insecticides, and was marketed only as a granular material, which was intended to be implanted into the soil. Aldicarb was then absorbed systemically and transported to all parts of the plant. Insects biting leaves months after an application could be annihilated. It was effective against nematodes, worms which destroy plant roots. It also had a mysterious quality of making plants look healthier, which was noted by many but never explained. Unfortunately, it was toxic to insects that are recognized as having beneficial effects on crops as well as those that destroy crops. Residues of aldicarb and its breakdown products began to appear in water supplies. Then in July of 1985, a group of Californians became ill after consuming watermelon containing aldicarb at a holiday picnic. Out of over one thousand complaints, more than 600 were probable and more than 300 were possible cases of poisonings. Fortunately, there were only two or three life-threatening cases. It is suspected that the contamination arose from an illegal use of aldicarb on a crop for which it had not been approved.

Resistance, particularly to insecticides, but to some extent among all classes of pesticides, is recognized as a significant problem. The use of pesticides results inevitably in a selection process for resistance. In general terms, the most common mechanisms for the development of resistance are either an increased capacity by the insects to biotransform insecticides to inactive products or to alter their target sites to prevent insecticide from binding to them. It is increasingly clear that insecticides are not a complete solution, and the trend today is toward systems of integrated pest management and insecticides in combination with other biologic techniques. Even resistance, however, can be exploited. Considerable research effort is being directed toward the genetic engineering of herbicide-resistant food plants. Resistant food plants could improve the efficiency of herbicides to kill weeds among the crops, and more specific chemicals could be directed against the weeds. This could reduce the need for tillage, which in turn would reduce erosion and energy, water, and fertilizer use.

The insecticides of the future are apt to be very different from those in use today. Hopefully, they will be safer for humans and cause fewer environmental problems, and they may be more biologic than chemical. For example, *Bacillus thuringiensis* is a bacteria that was found in dying silkworm larvae. It has been used for years as a microbial insecticide when applied as dried spores. If ingested by insects, the spores are activated to attack the gastrointestinal tract, and the insect stops feeding. The spores are virtually nontoxic to humans. This product is also unusual because it must be ingested to be effective. Most other insecticides are "contact poisons," and their major route of access to the insect is through its outer covering.

Insect sex attractants or pheromones are being used to lure insects into traps where they can be exterminated or exposed to chemosterilants. The latter chemicals are often highly toxic to humans, but they can be used safely in this way in a confined and controlled area. When the sterile insects are subsequently released into the environment, any mating efforts will be fruitless.

Cockroaches, beetles, and related insects have a tough outer shell (exoskeleton or cuticle) made of a material called chitin which is a complex sugar polymer. Contact insecticides readily penetrate through the cuticle. As insects develop from the worm-like larval stage, they undergo several molts in which they shed old cuticle and form a new one. There is considerable interest in finding effective inhibitors of chitin synthesis. If cuticle formation could be prevented, the larvae would be more vulnerable to the external environment and perhaps also have impaired reproductive capability. Such chemicals might be highly specific for some insects, and, therefore, may be safer for animals and humans. The approach is analogous to the action of penicillin, which kills bacteria by interfering with their ability to form cell walls.

Considerable effort is also being expended in attempts to identify insect growth hormones so that synthetic analogs or inhibitors might be designed for use as new insecticides. Such a chemical may one day be called the 2,4-D of insecticides.

SUMMARY POINTS

Pesticides are chemicals that are commercially exploited for toxicity and species specificity; they include insecticides, herbicides, rodenticides, and fungicides.

Methyl mercury, a widely used fungicide, has caused several devastating epidemics of human poisoning because of ignorance or a lack of ecological foresight.

SUMMARY POINTS (continued)

Warfarin, a widely used rodenticide, is at the same time a useful drug in humans to prevent excessive or inappropriate blood clot formation.

Selective, systemic herbicides kill broad leaf weeds; they are absorbed and translocated within plants to interfere with photosynthesis or promote lethal overgrowth.

Insecticides generate the largest sales volumes; many are descendants of chemical warfare agents known as nerve gases.

Resistance to pesticides, but especially insecticides, is a major limitation to their use; it can be countered partly by biologic pest control, crop rotation, and other measures.

BIBLIOGRAPHY

Baki F, Damluji SF, Amin-Zaki, L, et al.: Methyl mercury poisoning in Iraq. Science 181:230, 1973.

Canby TY: The rat, lap dog of the devil. National Geographic, July, 1977.

Darwin C: The Power of Movement in Plants. New York, D Appleton and Co, 1881.

Pierce PE, Thompson WH, Likoski WH, et al.: Alkyl mercury poisoning in humans. JAMA 220:1439, 1972.

Richardson DE: The isolation and identification of toxic coumarins. In Kadis S, Ciegler A, Ail SJ (eds): Microbial Toxins. VII: Fungal Toxins. New York, Academic Press, 1972.

Smith WE, Smith AM: Minamata. New York, Rinehart and Winston, 1975.

10

CHAPTER

THE RISE AND FALL OF DDT

The most effective insecticides ever synthesized were the so-called halogenated hydrocarbon or "hard-type," as typified by the first member of the group to be discovered, DDT. It and the dozens of related chemicals that were at one time on the American market have almost all now been banned. American scientists had a peculiar blind spot when it came to recognition of the problems that were being created by DDT. These problems were not unique to DDT. They had been encountered before DDT with chemicals that had similar properties but were not insecticides; they were encountered again with other chemicals that were introduced after DDT. For reasons that are not entirely clear it appears in hindsight that we have been slow learners.

The initials, DDT, stand for dichlorodiphenyltrichloroethane, a molecule that is packed with five atoms of chlorine. Fluorine, chlorine, bromine, and iodine are called the halogen family of elements, hence the name, halogenated hydrocarbon insecticides (HHI). In addition to the insecticides, halogenated hydrocarbons of many diverse chemical structures constitute perhaps the most serious problems in environmental pollution. The DDT molecule was first synthesized in 1874, but 65 years passed before a commercial use was found for it. Its insecticidal activity was first recognized by Paul Herman Mueller in 1939, and in 1948 he received the Nobel Prize in Medicine and Physiology for that discovery. In retrospect, it may be difficult to see why the selection committee regarded that as such an important contribution.

The discovery of the insecticidal activity of DDT coincided in time with the onset of World War II. There are those who believe that the impact of DDT on the eventual outcome of the war was second only to the development of the atomic bomb. In all previous wars in history more lives had been lost among combatants to communicable diseases than to actual battle injuries. Because DDT was not available to allied troops until late in the war, the same was true then, but an obvious turnaround was apparent by January of 1944. It was the first and only highly effective insecticide that was safe enough to be applied directly to human skin. When General Mark Clark's troops entered Naples, they dusted the remaining population with DDT powder in an attempt to avert an almost certain epidemic of louse-borne typhus. It became the first winter in recorded history that Naples did not have an outbreak, and the first time in history that an incipient epidemic of any communicable disease had been halted in its tracks.

DDT was also highly effective against fleas, which transmit bubonic plague, the most feared infectious disease in the world. It is still without peer in the control of mosquitos that carry malaria, which in turn is the most widespread infectious disease on earth. Some strains of mosquitos also transmit yellow fever. The latter two diseases have been virtually eradicated in the United States through the use of DDT and the efforts of the Army Corps of Engineers to drain swampy mosquito breeding grounds. DDT was also an agricultural insecticide par excellence. The use of it against only three insect pests: the Colorado potato beetle, the gypsy moth, and the bollweevil, saved untold millions of dollars in what would otherwise have been economic losses in food, forests, and cotton. The overall impact of DDT on human health and well being has been seriously compared with that of penicillin.

The physical and chemical properties of DDT and its many chemical relatives were critical to their effectiveness as insecticides. The only common structural feature among this group is their high degree of chlorination. From one- to two-thirds of their total molecular weight is made up of chlorine atoms. This makes them very insoluble in water, but they are soluble in organic solvents such as kerosene. Kerosene became a common vehicle for DDT sprays, and a not at all "inert ingredient" from the standpoint of human toxicity. Several early alleged cases of human DDT poisoning are recognized now to have been caused by the kerosene and not to DDT. This case of mistaken identity probably delayed the recognition of proper methods for treating poisonings that were in fact caused by HHI. Other so-called inert ingredients used as solvents or vehicles for pesticides that have been responsible for human morbidity or mortality include methylene chloride and polyvinyl acetate.

Low water solubility was an advantage to the use of DDT as an insecticide because once it had been applied to crops, it was not easily washed off by rain or watering. Its persistence decreased the number of times that it had to be reapplied during a growing season, which translated directly into decreases in costs to the farmer. As noted in Chapter 9, it is poorly absorbed through the intact human skin, but it penetrates through the cuticle or exoskeleton of insects. In order to be effective, DDT must be sprayed directly on the insect or else the insect must come into contact with the material on sprayed plants. Little if any of DDT's insecticidal activity can be ascribed to actual consumption of the chemical. Absorption through the cuticle, however, does take some time so that the spray operator is denied the vicarious pleasure of seeing instant convulsive death throes, but a satisfying number of carcasses is often found the next morning. Insects have a larger body surface to volume ratio than animals or humans so that they would receive a larger systemic dose from equivalent skin contact even without the difference in the rate of absorption.

The basis for DDT's insecticidal activity and for its toxicity in most mammalian species is an intense excitation of the brain and spinal cord (together these are called the central nervous system or CNS) which appears first as tremors and may progress to fatal convulsions. Convulsions are dangerous because it is not possible to breathe during seizures; thus, when the convulsions are continuous, the animal or human victim may die of suffocation. The appearance of convulsions following exposure to any chemical invariably indicates a dysfunction of the CNS. If the CNS is truly the target organ for DDT, then it should be found there during the time that an animal is showing signs of poisoning, and its concentration should be declining during the period in which the animal is recovering.

In the experiment summarized in Figure 10–1, rats were given a single toxic dose of DDT, and the times to the appearance of various signs of poisoning and recovery were noted. Some of the poisoned animals were sacrificed at various times to collect brain and fat samples for DDT analyses. As expected, the signs of poisoning became more severe as the brain concentration of DDT rose with time after dosing. Fatally poisoned animals died in the 8 to 12 hour period after dosing, which coincided with the peak brain concentrations of DDT. No deaths occurred after 12 hours, at which time the brain concentrations of DDT began to decrease. Surviving rats showed no toxic signs at 24 hours after dosing.

The levels of DDT in fat also rose with time after dosing (Fig. 10–1). The DDT in fat, however, could not possibly be responsible for toxic signs that must be mediated through the CNS. Note that the fat levels rise or remain constant even after complete recovery. Because the concentration in fat is rising at a time that the concentration in the brain is falling, the patterns suggest that DDT is being redistributed from brain to fat. This redistribution is in accordance with both the known high fat solubility of DDT, and the normal relative blood flows, which are high to brain and low to fat. Because the fat stores of DDT are biologically inert, this redistribution represents a mechanism for the detoxification of DDT. Indeed, for HHI as a class, redistribution to fat is the most rapid and efficient means for its temporary inactivation.

The permanent mechanism for the detoxification of DDT, as for most chemicals, is biotransformation to more water-soluble and less toxic forms followed by excretion of those forms in the urine. The metabolism of DDT, however, is unusually and painfully slow. Residues of DDT in fat persist for days or weeks after single doses. One of the products of DDT metabolism, called DDE, is retained

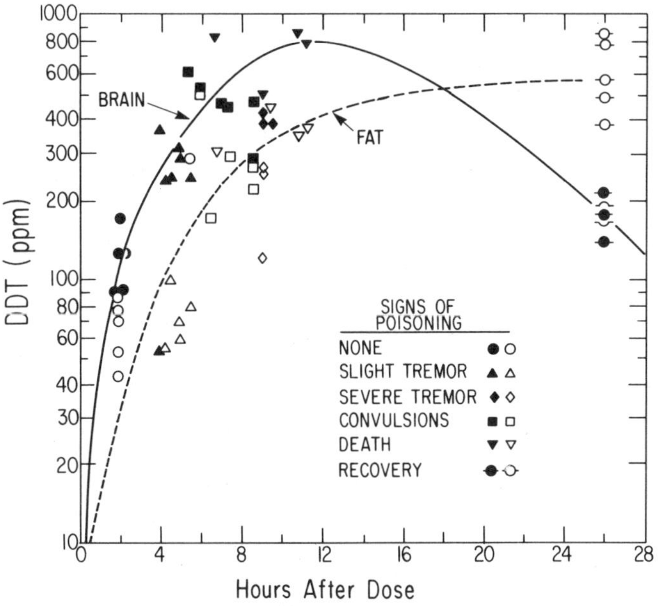

FIGURE 10–1. The concentrations of DDT in the brains (solid symbols) and fat (open symbols) of rats at various times after the administration of a single toxic dose. Different symbols indicate the signs exhibited by the animals at the time the samples were taken. (See text for additional details.) (From Dale WE, Gaines TB, Hayes WJ Jr, Pearce GW: Poisoning by DDT: relation between clinical signs and concentration in rat brains. Science 142:1474, 1963.)

in body fat even more tenaciously than is DDT itself. Therefore, the fat concentration of DDE is usually greater than that not only of DDT, but of all other DDT-derived materials combined.

The United States Public Health Service (USPHS) recognized the phenomenon of fat storage of DDT in humans as early as 1950. In persons not occupationally exposed to DDT, the bulk of those fat stores was believed to have had a dietary origin, namely, residues of DDT remaining on food crops, but especially from residues contained in animal fats. Extensive surveys were conducted by the USPHS in 1954–1956 and in 1961–1962 to measure the DDT content of human body fat. The USPHS results were confirmed by several independent laboratories.

In the 5 to 8 years between the two USPHS surveys the American production of DDT increased by about 40%. Still, the results of the surveys showed clearly that there had not been a parallel increase in the levels of DDT in fat in the general population. It had to be concluded that the American population had existed in a balanced steady state with respect to DDT intake, storage in fat, and excretion for as long as 5 to 8 years and perhaps for even longer. Those levels on the average were 5 ppm DDT or 12 ppm all DDT-derived materials.

The toxicologic principle illustrated by these findings is that continued or even increased intake of some chemicals within certain limits does not necessarily lead to directly proportional increases in accumulation or storage in the body. If intake is increasing and storage remains constant, it follows that excretion must also increase. An analogy can be made with attempts to fill a pail with a hole in its bottom from a water tap. If the leak is of a sufficient size, at low rates of water flow the water will leak out as fast as it comes in, and none will accumulate in the pail. If the tap is opened further, at some point the rate of flow in will equal and begin to exceed the rate of flow out, and water will slowly accumulate in the pail. Further increases in flow rate from the tap will lead to increasingly rapid filling of the pail. When the tap is turned off, the pail will eventually drain completely. As the level of water in the pail rises, it will be forced out of the hole faster because of the increased pressure. There is also an analogy to that phenomenon in humans in that the size of the leak (excretion) is also increasing so that a steady state is reached in which the pail is neither full nor empty. During the 1950s and 60s the general population of the United States appeared to be in such a steady state, and we were consuming about 30 μg of DDT (60 μg of all DDT-derived material each day) along with our daily bread.

It is important to distinguish between a steady state and saturation of fat stores. In the pail analogy saturation occurs when the pail is full and it can hold no more. Everything that flows in thereafter immediately overflows. The saturating concentration of DDT in human body fat is unknown, but estimates from animal studies suggest that it is much higher than the measured steady state levels. It is revealing to calculate the total body burden of DDT in the average adult from the steady state levels in fat. Assuming that the body is 20% fat, and that the average adult weighs 70 kg, it follows that:

$$70 \text{ kg} \times 0.2 = 14 \text{ kg of fat}$$

$$14 \text{ kg fat} \times 12 \text{ mg DDT-derived material/kg fat} = 168 \text{ mg residue}$$

$$168 \text{ mg residue}/70 \text{ kg} = 2 \text{ mg residue/kg body weight}$$

From animal experiments and some limited clinical experience in human poisonings, it is thought that a single dose of DDT that frequently causes convulsions is 20 mg/kg, and the mean lethal dose of DDT in an adult human is about 200 mg/kg. Thus, with respect to the average total body burden, the safety factor for serious and lethal poisoning was 10 and 100, respectively. That body burden, however, was presumably safely stored in body fat where it could do little or no harm. Indeed, that body burden has yet to be associated even in retrospect with any human illness. It is possible that such a body burden could have been tolerated indefinitely. Certainly, much higher fat levels appeared to be well tolerated in individuals occupationally exposed by manufacturing, distributing, and applying DDT. Human volunteers have consumed DDT in "spiked" meals at rates that were more than 1000 times the average daily intake without obvious harmful effects. A total of 24 subjects consumed 35 mg/kg per day for almost two years, and their fat levels ranged from 100 to 600 ppm.

Because DDT was without peer as an insecticide and because no evidence existed for adverse health effects at the levels to which the population was being exposed, it might be with us still were it not for a series of findings and considerations, each relatively minor in itself, but which *in toto* made some ominous predictions for the future. Animal experimentation pointed out what was obvious from the pail of water analogy, namely that the rate of intake could not continue to rise indefinitely without a departure from the steady state fat storage levels. Figure 10–2 illustrates an intensive feeding experiment with rats in which the diet contained DDE, but the same phenomenon could have been shown with DDT. The fat concentrations rose until they reached steady state levels after 90 to 100 days. The fat levels plateaued at about

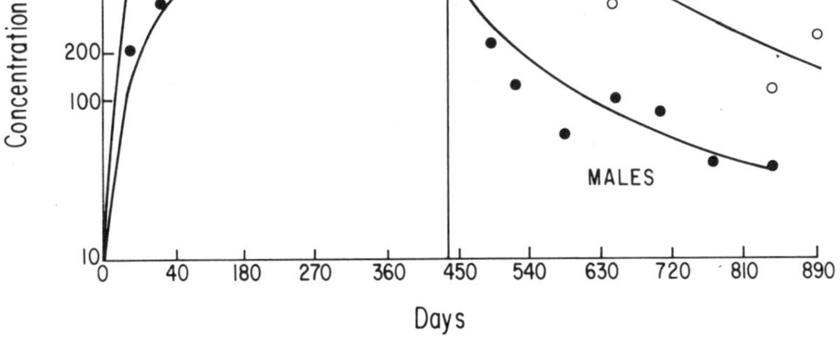

FIGURE 10–2. Storage of DDE in the body fat of male and female rats during and after feeding a diet containing 200 ppm. (After Durham WF: Body burden of pesticides in man. Ann NY Acad Sci 160:183, 1969.)

5000 ppm for female rats and at about 1000 ppm for males, illustrating yet another sex difference. The animals exhibited no adverse effects, but when plain diet was substituted for the one containing DDE, it took a long time for the fat levels to fall to prefeeding concentrations. Measurable amounts of DDE were still present more than a year later.

A more alarming variation on this experiment showed that DDT as stored to these levels in body fat represented something of a potential time bomb. The experiment started out in the same way as above except that the rats were fed DDT to very high, but nontoxic, steady state body fat levels. After the steady state was reached, the rats were rationed to only half of their normal daily food intake, but the diet still contained the same concentration of DDT. Within a few days many animals were exhibiting tremors and some convulsed and died. The levels of DDT in the brains of these fatally poisoned animals were the same as the brain levels at the peak in the experiments of Figure 10–1, in which the animals were given only a single dose of DDT. During starvation, body fat is preferentially consumed as fuel to sustain life. As the body fat deposits decrease in size, they begin to release stored DDT into the blood. As the concentration in blood rises so also does the concentration in all other tissues in the body, including the brain. In these starved rats the DDT concentrations quickly reached toxic levels even though their daily intake of DDT had been cut in half. These findings gave rise to grave legitimate concerns about the use of DDT in developing nations where there is a great need for insecticides and starvation may be endemic.

Although DDT was the most widely used insecticide in the HHI class, dozens of other closely related chemicals were also on the American market that had the same physicochemical properties and the same toxic effects. Others closely related to DDT included methoxychlor, dilan, and Kelthane. The cyclodiene subgroup included aldrin, dieldrin, and endrin. Miscellaneous HHI included lindane, toxaphene, mirex, and Kepone. Some of these were more acutely toxic than DDT, and some less. Although almost all are now banned in the United States, during the 1950s and 1960s all were being used.

In the surveys of body fat concentrations it was common to measure DDT by an instrument called a gas chromatograph. This machine records the DDT concentrations in appropriately prepared samples as a tracing in which the heights of the peaks are proportional to the amounts of DDT. The detection device, however, was not specific for DDT, and it would also detect and record any similar chemicals in body fat. The metabolic products of DDT, for example DDE, would show up as separate peaks. It soon became obvious that new and strange peaks were being recorded that could not have been derived from DDT. "Peak counting" became a serious preoccupation for some toxicologists. These peaks were slowly and tediously identified as belonging to other HHI and their metabolic products, which were also being stored in human body fat. Because all HHI appear to act by the same basic mechanism, they would be predicted to have at least additive toxicity to that of DDT. The issue of the fat concentration of DDT was suddenly clouded because of the other HHI that were also present. The total body burden of DDT became less relevant than the total body burden of all HHI.

Feeding studies in rats showed that there could be important and unpredictable interactions with respect to fat storage when two or more HHI were fed at the same time. When both DDT and dieldrin were fed together, the concentrations of dieldrin in body fat were lower than those in rats fed only dieldrin at the same level, whereas the fat concentration of DDT was higher than that in rats fed only DDT. Thus, the issue of the daily intake also became clouded because of such interactions with respect to fat storage.

Another alarming discovery was that DDT and other HHI were able to cross the human female placenta, which serves as only a partial protective barrier for the fetus. These materials were identified in the organs and tissues of stillborn infants although there was no reason whatever to believe that they were responsible in any way for the stillbirths. Ironically, the concentrations of DDT found in random samples of human female breast milk were 2 to 6 times higher than that permitted in cow's milk for sale to the general public. A wry ecology poster of the day showed a human female breast with the legends, "Caution, keep out of the hands of children" and "Milk in such containers may be unfit for human consumption."

Two effects of DDT have been ascribed to environmental levels of exposure. The first must have been common, but it is not generally regarded as an adverse effect. The induction of hepatic microsomal enzymes by almost all HHI is easily demonstrable in rats. It is known that certain barbiturate drugs are inactivated in rats by those liver enzymes. If the drug is injected in an appropriate dose, after a few minutes the rat will go to sleep. At the same time those liver enzymes are in the process of converting active drug to inactive products. As the level of active drug in the blood falls, the active drug in the brain, which is responsible for the anesthetic effect, tends to move into the blood, and the rat will awaken. The length of time that the rat remained asleep is an indication of the degree of activity of its hepatic microsomal enzymes. If the enzymes are inhibited the sleeping time will be longer than normal because the drug is inactivated more slowly. Conversely, if more enzymes have been induced by exposure to

HHI, the rat will sleep for a shorter length of time than animals not so exposed.

If one performs a feeding experiment of the type shown in Figure 10–2 using DDT and measures the fat levels at the steady state, an appropriate concentration in the diet will result in 16 mg DDT + DDE/kg of fat. In rats not fed DDT, the levels were 2 mg/kg of fat. The levels in control animals reflect background environmental contamination of their feed; it simply was not possible at the time to buy rat feed that was completely free of DDT. The control rats slept for an average of 81 minutes after the barbiturate was injected, whereas the animals fed DDT slept only half as long. Because the average steady state fat concentration in the human population was 12 mg/kg body fat, we may infer that in all likelihood some degree of hepatic microsomal enzyme induction was occurring in humans as well. No evidence exists to suggest that the inferred phenomenon was adversely affecting human health. Nor does there seem to be any hint that commonly used drugs that are inactivated by microsomal enzymes were noted to be less effective during that period. A carefully controlled study, however, might have revealed subtle changes. Induction of microsomal enzymes accounts for at least some of the interactions among HHI with respect to fat storage, and it may have played a role in the maintenance of steady state fat levels in the face of increased intake, i.e., induction of its own metabolism may have been one of the means for hastening its excretion.

The other possible effect of DDT is dangerous, but also rare. Although it has not been conclusively proven for any HHI, there have been isolated reports of an association between intensive exposure to chlordane and lindane and serious damage to the bone marrow (aplastic anemia). Because the HHI have so many other properties in common, all, including DDT, must be suspected of producing this effect, but rigorous proof is still lacking, and DDT is guilty only by association.

In the final analysis, the decision about DDT was made on environmental grounds. It was the first time in history that a case of that type was decided on grounds other than concerns about human health. The environmental movement was galvanized into action by a series of events, perhaps the first of which was the recognition of the staggering magnitude of the global contamination. In less than 30 years DDT had reached measurable levels in the remotest parts of our planet. It was identified in owls that never ventured south of the Arctic Circle and in Antarctic penguins. No body of water and no living species was free of DDT, as already alluded to by the problem with rat feed above. It was estimated in 1968 that one billion pounds were in the environment. Mankind had been undone by that valuable property of DDT as an insecticide, namely, its persistence.

Evidence was available not only to show that DDT was ubiquitously distributed among living species, but even more alarming was evidence to show that it was being concentrated as it was passed upward through natural food chains. Figure 10–3 illustrates a field study that was conducted around an estuary on Long Island. Specimens of various forms of plant and animal life were collected from the waters and along the shores and analyzed for DDT. The numbers under each listed species indicate the ppm of DDT found in each specimen. The chart is organized along known natural food chains with the higher trophic levels on the right. The carnivorous birds feed on fish, which eat crustacea/or plants. The lines connect species that feed mainly on the indicated lower forms.

In Figure 10–3, one can start with plankton, which contained 0.04 ppm DDT by wet weight, and follow any of the lines at ran-

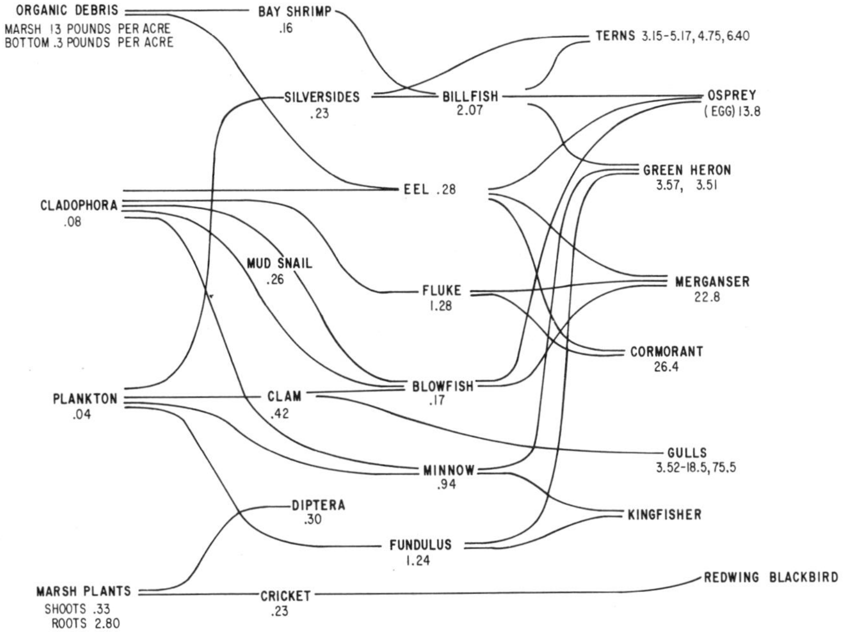

FIGURE 10–3. A natural food web of plants, crustacae, fish, and birds in a Long Island estuary, separated into trophic levels from left to right. Lines indicate the species that feed on specific members of lower levels. The numbers indicate the whole organism concentration of DDT in ppm wet weight. (From: Woodwell GM: Toxic substances and ecological recycling. Sci Am 216:24, 1967.)

dom. It can be seen that the concentration of DDT rises each time it enters a new trophic level. In the extreme case of a ring billed gull, the concentration has risen by 2000-fold to 75 ppm. The highest concentrations were found in scavenging and carnivorous fish and birds, but the birds had 10 to 100 times the concentrations found in fish. The levels found in some birds approached those known to be lethal from laboratory studies.

Concentrations of DDT in natural food webs occurs because of the persistence and chemical inertness of DDT. The process is shown diagrammatically in Figure 10–4. The areas of the individual rectangles represent the total biomass for each indicated trophic level. Arrows pointing to black dots outside of the biomass represent the small losses of DDT resulting from metabolism and excretion, or to the loss of individual members of that biomass.

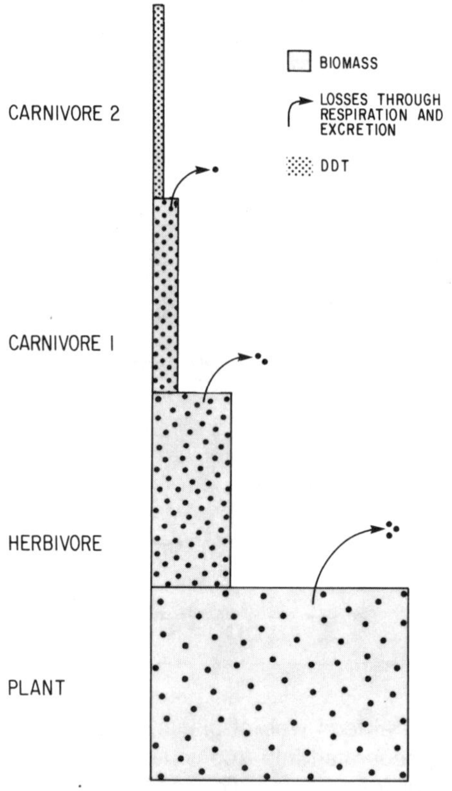

FIGURE 10–4. How DDT residues become concentrated as they are passed up a simple food web. (See text for additional details.) (From Woodwell GM: Toxic substances and ecological recycling. Sci Am 216:24, 1967.)

Clearly, these losses are very small relative to the total amount present in any biomass.

The plant biomass is very large. Although it is dwindling at an alarming rate, forest trees make up 90% of the earth's biomass. The plant biomass, however, contains relatively low concentrations of DDT. Assume that in passing upward from one trophic level to the next, half of the biomass consumed is physically incorporated as part of the permanent structure of the species in the higher level. The other half consumed is broken down, used as energy, and the waste products are excreted. Thus, the biomass decreases by one half each time it moves up a trophic level. Almost all of the DDT, however, is retained so that the decrease in biomass results in a twofold concentration of the DDT.

Any increase in environmental DDT, such as increased application, improper disposal, or an accidental spill would be passed directly up the food web. The carnivorous species at the top of the food web also tend to be long-lived. These, therefore, will accumulate DDT over a longer period of time, resulting in even higher concentrations toward the end of their lifespan. Humans, of course, are the longest lived carnivores on earth.

Birds played a prominent role in the downfall of DDT. The town of Hanover, NH, is situated directly across the Connecticut River from the neighboring community of Norwich, VT. Citizens in both communities were concerned about Dutch elm disease, which had destroyed many of their lovely trees. In the mid 1960s the town of Hanover mounted an intensive spraying operation with DDT to combat the disease, whereas the town of Norwich voted against spraying. The stage was set for a natural experiment. A group of scientists at Dartmouth College began a newspaper and radio campaign for citizens from both communities to bring in all dead or dying birds for DDT analyses. Over the next several months 151 dead birds were found in Hanover and only 10 in Norwich. The brain concentrations of DDT were clearly higher in the Hanover birds than in the Norwich birds. The town of Hanover then decided that it preferred its robins to its elm trees.

It is now well established that DDT exposure results in abnormally thin egg shells in some species of birds. The shells are so fragile that they break prematurely, and the embryo is lost. Such species as the peregrine falcon and the brown pelican of Southern California were very nearly exterminated.

Beginning in the 1960s a series of fish kills numbering in the millions occurred in the lower Mississippi basin. In 1964, these were found to be caused by minute traces of endrin. The lethal concentration of endrin for fish (LC50) was so low that techniques for actually measuring it were not available until 1961. Although it is hard to assess its total impact on the environmental movement, the appearance of Rachel Carson's book, *Silent Spring,* must have played some role. Like *Brave New World* and *1984, Silent Spring* was not very scientific and certainly exaggerated, but all were prophetic. In 1972, after a set of hearings on the issue of whether or not it should be banned for domestic use in the United States, William Ruckelshaus, the EPA administrator at the time, reached the conclusion that there was no way in which the use of DDT could be regulated that would adequately protect the environment. The banning of the cyclodiene group followed shortly thereafter.

SUMMARY POINTS

Paul Herman Mueller received a Nobel Prize for discovering the insecticidal activity of DDT; it was an agricultural insecticide without peer in forests, on potatoes, and on cotton.

DDT could control the insect vectors for louse-borne typhus, bubonic plague, malaria, and yellow fever; it appeared to be relatively nontoxic to humans and animals.

DDT produces convulsions in insects and animals when the brain concentration reaches appropriate levels, but its high lipid solubility leads to inert storage in body fat.

This storage in fat resulted in its concentration in natural food chains, where the highest concentrations were found in the long-lived carnivores, including humans.

Because it was chemically inert, man managed to contaminate the entire planet with DDT in less than 30 years; the peregrine falcon and the brown pelican faced extinction.

DDT use in this country was banned in 1972; it was the first time in history that a chemical had been banned on ecological grounds rather than human health concerns.

BIBLIOGRAPHY

Carson R: Silent Spring. Boston, Houghton Mifflin, 1962.

Dale WE, Gaines TB, Hayes WJ Jr: Storage and excretion of DDT in starved rats. Toxicol Appl Pharmacol 4:89, 1962.

Dale WE, Gaines TB, Hayes WJ Jr, Pearce GW: Poisoning by DDT: relation between clinical signs and concentration in rat brains. Science 142:1474, 1963.

Durham WF: Body burden of pesticides in man. Ann NY Acad Sci 160:183, 1969.

Woodwell GM: Toxic substances and ecological cycling. Sci Am 216:24, 1967.

Wurster CF Jr, Wurster DH, Strickland WN: Bird mortality after spraying for Dutch elm disease with DDT. Science 148:90, 1965.

11

CHAPTER

CHLORINES, BENZENE RINGS, AND DIOXINS

The purity of our foods, drugs, and cosmetics is regulated by law (see Chapter 8), but no similar laws regulate the purity of pesticides or other commercially important chemicals. Technical grade chemicals are usually the cheapest that one can buy because they contain the greatest amounts of impurities. Although the definitions of the terms may not be uniform from company to company, in terms of increasing purity chemicals may be classified as practical grade, purified, chemically pure, reagent grade, and so on. When a commercial chemical is deliberately manufactured in such a way that it contains many different but closely related active ingredients, the problem of impurities is magnified. Such commercial mixtures may contain so many individual chemicals that they defy precise description of even the desired ingredients, let alone any impurities that might have been introduced during the manufacturing process. Those impurities, however, may compound the problem of environmental pollution by the mixtures.

Few naturally occurring organic compounds have a carbon-to-chlorine bond, and it was a sad day for the environment when chemists learned how to make such compounds synthetically. Between 1941 and 1953, a period which overlaps the introduction of DDT in the United States to the general market, hundreds of thousands of cows died from a previously unknown disease. It was at first called virus X disease, but when infection was ruled out as a cause, it became X disease or hyperkeratosis. In affected cattle the skin became dry and wrinkled, and it had an unusually high content of keratin. Keratin is an insoluble protein, which makes up the structural components of skin; leather is almost pure keratin. Loss of hair was also common, and the hides from diseased cattle were spoiled for many commercial uses. In addition, cattle showed some signs of liver injury, and it was noted that the blood plasma concentrations of vitamin A were abnormally low. Because vitamin A supplementation in affected cows reversed some but not all of the signs of the disease, the condition was not simple vitamin A deficiency.

The cause of the disease was eventually traced to chemical contaminants in commercial cattle feed. The contaminants were a series of polychlorinated naphthalenes (PCN, Halowax). The chemical structure of one member of this group is shown in Figure 11–1. Chemicals containing the hexagonal ring structure (a benzene ring) are called aromatics. Commercial PCN mixtures, however, contained dozens of related chemicals having fewer chlorine substituents (homologs). When there are fewer chlorine substituents, it is possible for them to be in different positions on the naphthalene ring, which makes them chemically different compounds even when they have the same total number of chlorines (isomers). The isomers and homologs constitute fairly minor chemical variations around a central theme, but they do not necessarily represent minor toxicologic variations around a theme. Several different companies were manufacturing PCN, and any one company's product almost certainly had a different composition from that of any other company's product because of differences in conditions during the manufacturing process.

Indeed, manufacturing conditions within a given company were often changed deliberately in order to obtain mixtures of different compositions, and, therefore, also of different physicochemical properties. Different PCN mixtures were intended for different commercial uses such as synthetic waxes, lubricants, and electrical insulators. Because the mixtures could not be defined in precise chemical terms, it was common to describe them simply in terms of the average number of chlorines per naphthalene ring structure without regard to their location. Obviously, this is not a precise chemical definition, and even two products with the same average number of chlorines per naphthalene could have different chemical compositions, i.e., the average of 2 and 8 is the same as the average of 4 and 6.

The PCN mixtures had many properties in common with DDT. They were insoluble in water, but soluble in fat and some non-

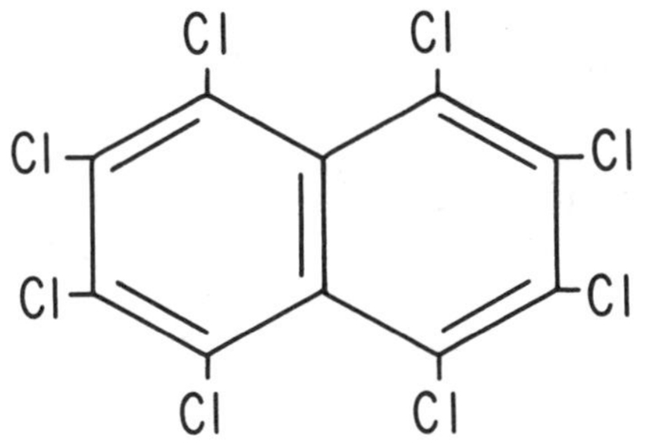

FIGURE 11–1. The fully chlorinated ingredient in polychlorinated naphthalene (PCN). The commercial mixtures of PCN contained not only the above ingredient, but dozens of related chemicals having fewer chlorines located in all possible positions shown around the ring structure. The mixtures probably contained numerous impurities as well.

aqueous solvents. They were chemically inert, and slowly biotransformed and excreted. Although proof positive is lacking, it may be safely inferred that they were persistent in the environment, stored in human body fat, excreted in breast milk, and concentrated in natural food webs. Thus, nursing calves contracted hyperkeratosis without consuming contaminated feed. The disease was also associated with other poorly defined commercial products, such as waste crankcase oil and wood preservatives, which may or may not have contained PCN. In fact, how the cattle feed came to be contaminated with PCN was never clarified. It might have been an accident or done with malicious intent, but in retrospect, it seems more likely that it was already present in some natural component that was used to make the feed.

Workers in the plants that made PCN often suffered from a common occupational dermatitis known as chloroacne. Chloroacne is characterized by severe skin eruptions closely resembling the disfiguring acne vulgaris of teenagers. An association between highly chlorinated benzene ring structures and a risk of chloroacne had been recognized since the turn of the century. Besides chloroacne, workers often had mild signs of liver disease.

Some of the individual pure chemical ingredients found in PCN mixtures were synthesized and tested for their ability to produce hyperkeratosis when they were applied to rabbit ears. This test had shown close correlation with the ability of various chemicals to produce chloroacne in humans. Those ingredients having three or fewer chlorines did not produce a reaction on rabbit ears, but the reaction was produced with increasing intensity as the number of chlorines increased thereafter.

Another group of somewhat related chemicals that were in use at about the same time, and that were also known to cause chloroacne among workers handling them, were the chlorophenols. Unlike the PCN, the chlorophenols were pesticides and used as fungicides, sanitizing agents, and wood preservatives. One member of this group, penta, was introduced in Chapter 9. A series of these were marketed under the trade names of Dowicides, and the largest volumes of sales were for the 2,4,6- and the 2,4,5-trichlorophenols (Fig. 11–2). 2,4,5-Trichlorophenol was used in large amounts as the starting material for the preparation of the herbicide, 2,4,5-T. Pure samples of various chlorophenols failed to produce hyperkeratosis when applied to rabbit ears, but technical grade samples did produce the reaction. Therefore, the ingredient responsible for hyperkeratosis or chloroacne appeared to be either an impurity introduced during the manufacturing process or a spontaneous breakdown product of one or more of the desired ingredients.

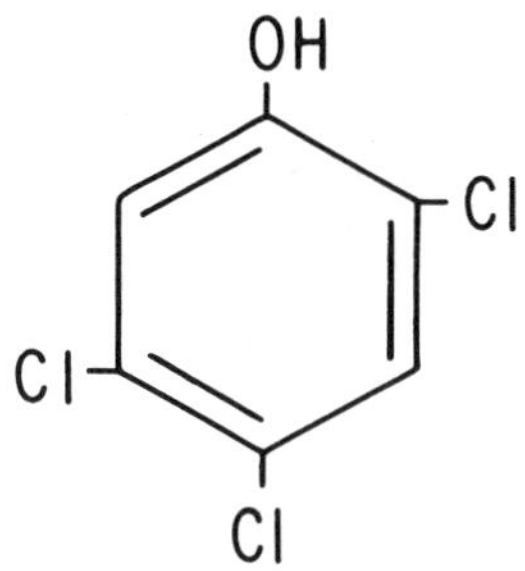

FIGURE 11–2. 2,4,5-Trichlorophenol or Dowicide 2, which is used as a fungicide, antimicrobial agent, and wood preservative.

Ever since 1929 another group of chemical products closely related to PCN had been in widespread use in the U.S. These were the polychlorinated biphenyls or PCB. Their physical and chemical properties were similar to those of both PCN and DDT. The biphenyl ring structure has ten possible positions for chlorine substitution (Fig. 11–3) so that PCB mixtures contain even more ingredients than PCN. There are 209 possible homologs and isomers of PCB. On chemical grounds, however, certain ingredients are more likely to be generated in substantial amounts than others. American PCB products are said to contain significant amounts of only about 50 distinct compounds. In 1970 the U.S. domestic sales of PCB were 34,000 tons, as compared with 70,000 tons of DDT.

Some PCB products were mobile liquids and were used, for example, as microscope immersion oils, others were hard resins and even crystalline-like solids. Certain mixtures were extremely heat stable and were useful in fireproofing; others were used as

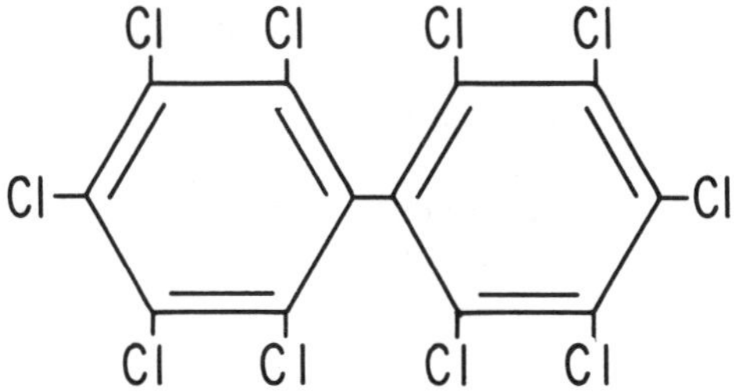

FIGURE 11–3. The fully chlorinated ingredient in polychlorinated biphenyl (PCB). The commercial mixtures of PCB contained not only the above ingredient, but dozens of related chemicals having fewer chlorines located in all possible positions shown around the ring structure. The mixtures contained numerous impurities as well.

plasticizers, lubricants, adhesives, ingredients in paints and varnishes, and insulators and coolants in electrical transformers. These products were also defined on the basis of the average number of chlorines present on the ring structure, with ten chlorines representing 100% saturation. Araclor 1260 has six chlorines (60% saturation) on the average on each ring structure.

These products were on the market 10 to 15 years before DDT was generally available. Their total sales may have been smaller than DDT, but it was not until the 1970s that they began to receive any attention as a possible environmental problem. Why were they overlooked for so long? At least part of the problem was that DDT and related hard-type insecticides were getting all the attention. Each development in that area seemed to unveil a new threat, and those chemicals were known to be toxic to nontarget organisms, whereas PCB had not been extensively studied. Another reason was the fact that the physical and chemical properties of PCB were similar to those of the HHI. When fat samples were extracted and analyzed for HHI by gas chromatography, PCBs were also extracted and detected. This was a period in which new HHI and metabolites of them were regularly being discovered in fat. When new unidentified peaks showed up in recordings, it was assumed that they were unidentified HHI or their metabolites. Indeed, the properties of HHI and PCB were so similar that the eventual separation and positive identification of PCB presented a formidable analytical problem, as was that of separating and identifying the individual desired components in PCB mixtures from their impurities. With 50 or more individual chemical ingredients, the gas chromatographic tracings resembled insane picket fences.

By the time PCB were identified in various biologic samples, the environmental contamination was as bad as it had been at the heighth of DDT use. Again, every major river system in the U.S. was contaminated. The Hudson River was so badly polluted by PCBs from discarded electrical transformers along its banks that it was closed to commercial fishing for a period in 1976. Chemists found PCB in some fish in concentrations that were 100 times those of DDT.

Both DDT and PCB have significant vapor pressures, and they tend to codistill with water. Evaporation of contaminated surface waters is, therefore, accompanied by an increase in atmospheric PCB and DDT, which is one means for their spread throughout the world. Under the influence of ultraviolet radiation atmospheric DDT can be converted to recognizable PCB, and PCB vapors have been shown to be converted to chlorophenols.

From the beginning, toxicologists were bothered by wide variations in the toxicity from one PCB mixture to another. A series of Araclors ranging from an average of 20 to 60% saturation had acute oral LD50 values in rats ranging from 4 to 19 gm/kg. Thus, PCB are considerably less toxic acutely than DDT, and they produced CNS depression instead of excitation. Other PCB products, however, were more toxic than the Araclors and all products had a high tendency to produce chronic toxicity. Poisoned animals had signs of liver damage, like the lesions produced by PCN. An unusual effect was noted in poisoned birds. The hearts of both birds and mammals are enclosed in a two-layer sack called the pericardium. In some rare diseases fluid accumulates in this sack, and the condition is called hydropericardium. Because PCB regularly produced hydropericardium in birds the unknown responsible ingredient was called the chick edema factor.

Starting in 1968, a new skin disease began appearing in the western prefectures of Japan. Eventually, about 1000 people were affected. The signs included severe acne-like skin eruptions, jaundice and other signs of liver damage, and an abnormal pigmentation of the skin and nails similar to that seen with HCB (see Chapter 9). The disease was finally associated with the use of a particular brand of rice oil for cooking, and it became known as Yusho disease. An analysis of a suspect sample of the oil revealed the presence of 2000 ppm PCB. One feature of the complexity of the ingredients in PCB products is that no two consumer products are exactly alike. Thus, the distributions of the various ingredients can be used precisely to "fingerprint" various products. The distribution of PCB in Yusho oil corresponded to a product called Kanechlor which was about 48% saturated with chlorine. An investigation at the rice oil factory revealed that Kanechlor was being used as a heat transfer agent in some of the machinery, and that an unsuspected leak was resulting in contamination of the oil.

Ten abnormal or stillborn infants were delivered from mothers known to have consumed contaminated oil, suggesting that PCB may be teratogenic. Despite intensive therapy in many forms, the signs of Yusho disease persisted in some patients for as long as 2 years after termination of the exposure. The manufacturing of PCB in the U.S. was halted in 1979, seven years after the banning of DDT.

In 1957, the poultry broiler industry in the U.S. suffered a severe financial blow when about half of the existing flocks died from a previously unknown disease. The chickens were droopy, had ruffled feathers, apparent breathing problems, signs of liver damage, and hydropericardium, as noted with some samples of PCB, hence the name chick edema disease. The cause of the disease (chick edema factor) was traced to the lipid fraction of some brands of commercial chicken feed. This fraction contained various vegetable oils and animal fats, and it was called "toxic fat." When its use was banned, chick edema disease disappeared.

Similarities in the names, physical appearance, and packaging of consumer products invites confusion and sometimes disaster. A chemical company in Michigan made both a fire retardant product (Firemaster) which was toxic, and a nutrient additive for cattle feed (Nutrimaster), which was essentially nontoxic. In 1973 these were accidentally confused, and Firemaster was incorporated into feed pellets. Firemaster consisted primarily of polybrominated biphenyls (PBB), which had the same basic chemical structure of PCB (Fig. 11–3), except that bromines were substituted for chlorines. Some feed pellets contained as much as 13,500 ppm PBB, and livestock were exposed for a minimum of 9 months. The first illnesses and deaths of livestock were noted within 3 to 4 months. Over the next three years 32,000 cows, 6000 pigs, 1400 sheep, and 1.5 million chickens died or were destroyed because of contamination. Also destroyed were huge amounts of eggs, cheese, butter, and dried milk. *Time* magazine referred to the incident as "cattlegate" because of alleged attempts at coverup by state officials. Epidemiologic studies of randomly selected groups from quarantined and nonquarantined farms showed 37% of the individuals examined had neurologic abnormalities and 20% had chloroacne. A year after PBB had been removed from the food supply, the majority of a large sample of exposed Michigan residents had detectable levels of PBB in their blood serum and body fat.

Throughout much of the 1960s, the U.S. engaged in extensive defoliation programs in the Vietnam conflict. The publicly stated purpose was to decrease cover and concealment for enemy forces, but it is likely that food denial and the creation of a flow of refugees from Communist controlled areas to areas controlled by South Vietnam were covert purposes. Herbicides were sprayed from large C-130 cargo planes in concentrations 3 to 10 times higher than those recommended for good agricultural practice. The spraying was halted in 1970, but for years after crisscross spray patterns were clearly visible from the air. Some mangrove forests had been reduced to bamboo thickets. The most extensively used product was known by the code name of Agent Orange. It consisted of about equal parts of 2,4-D and 2,4,5-T in the form of their n-butyl esters.

At that time, not much was known about either the acute or the chronic toxicity of these auxin-type herbicides except that they seemed to be only moderately toxic to laboratory mammals. Because 20,000 tons of Agent Orange were eventually to be sprayed over Vietnam, it seemed desirable to intensify research into their possible toxic effects. To the great horror of the U.S. producers of these herbicides, several studies quickly indicated that at least some samples of 2,4,5-T were teratogenic (see Chapter 20) and produced cleft palettes and kidney malformations in animals. Because this information came almost on the heels of the thalidomide disaster in Europe (see Chapter 20), both the scientific and the lay communities were up in arms.

All of the above seemingly unrelated and disparate threads of calamities began to come together in 1967 when a team from Proctor and Gamble isolated and identified the chick edema factor from carefully preserved samples of toxic fat. The factor belonged to a group of 75 chemicals, of which the most toxic was 2,3,7,8-tetrachlorodibenzo-p-dioxin (TCDD or "dioxin"). A few years later, scientists from the Dow Chemical Company showed that the samples of 2,4,5-T that were teratogenic were contaminated with dioxins and a related group of chemicals, the dibenzofurans (Fig. 11–4).

Once you know what to look for, things become considerably easier to find. Table 11–1 shows that either dioxins or dibenzofurans (DBF) or both are contaminants in at least some samples of four of the products associated with the problems enumerated above. Besides toxic fat and 2,4,5-T, they have been identified in PCB and some chlorophenols. Note the bizarre circumstance that the old PCN are also found as contaminants in some PCB.

In the same way one can also summarize the animal toxicology and the human experience, as shown in Table 11–2 which lists the various products associated with a given effect. The considerable overlap in the products and the effects they produced makes a strong case for a common or at least similar cause. PCN and PCB were each associated with four of the five effects, although not the same four. PBB and 2,4,5-T are each associated with three of the five effects. A "bromo-dioxin," however, has yet to be identified in PBB. It is possible that some of these effects are caused exclusively by the dioxin or DBF contaminants, whereas other effects may be caused by the intended ingredients of the products.

An additional effect that could have been added to Table 11–2 is "induction of hepatic microsomal enzymes" because dioxins and DBF are among the most potent inducing agents known. In fact, TCDD is 30,000 times more potent than other commonly used

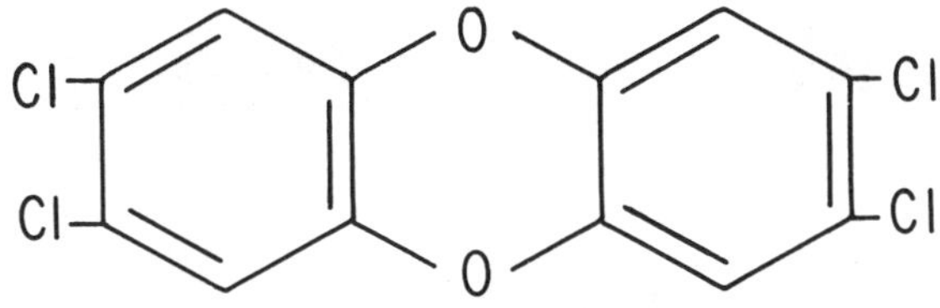

FIGURE 11–4. The most toxic of the 75 known dioxins, 2,3,7,9-TCDD. A related group of chemicals, the dibenzofurans, lacks one of the oxygen (O) bridges in the center ring so that the two benzene rings are joined directly at that site, e.g., 2,3,7,8-TDBF.

TABLE 11–1. TOXIC CONTAMINANTS IN AROMATIC CHLORINATED HYDROCARBON MIXTURES, PRODUCTS, AND TOXIC FAT

Substance	Toxic Contaminant
2,4,5-T	Dioxins, DBF
PCB	DBF, PCN
Chlorophenols	Dioxins, DBF
Toxic fat	Dioxins

(After Kimbrough RD: Toxicity of chlorinated hydrocarbons and related compounds, a review including chlorinated dibenzodioxins and chlorinated dibenzofurans. Arch Pathol 94:125, 1972.)

inducers in laboratory animals, and the effect is much longer lasting (months as compared with a week or two). Sometimes disturbances in heme metabolism are associated with the induction process because cytochrome P450 is a heme-containing enzyme. The disturbance in heme metabolism may take the form of porphyria cutanea tarda as with HCB exposure (see Chapter 9), perhaps accounting for the hyperpigmentation of the skin in Yusho disease.

Although prominent species differences exist (see Chapter 3), in some species TCDD is the most toxic and teratogenic synthetic chemical known to man. The acute oral LD50 in the guinea pig is 1 μg/kg. Even in the most resistant species, however, it is still in the super toxic range. The cause of death is still uncertain in the guinea pig, but liver damage is probably responsible for death in mice, rats, and rabbits. Its physicochemical properties are also remarkable because they are similar to the HHI and the major ingredients of the products in which they are found. Instead of being highly reactive and unstable as are many highly toxic chem-

TABLE 11–2. EFFECTS OF EXPOSURE TO VARIOUS AROMATIC CHLORINATED HYDROCARBON MIXTURES, PRODUCTS, AND TOXIC FAT

Substance	Biologic Effect
PCN	Chloroacne, liver damage, X disease of cattle, chick edema
PCB	Chloroacne, liver damage, teratogenesis, chick edema
2,4,5-T	Chloroacne, liver damage, teratogenesis
PBB	Chloroacne, liver damage, teratogenesis
Chlorophenols	Chloroacne, liver damage
Toxic fat	Teratogenesis, chick edema

(After Kimbrough RD: Toxicity of chlorinated hydrocarbons and related compounds, a review including chlorinated dibenzodioxins and chlorinated dibenzofurans. Arch Pathol 94:125, 1972.)

icals, dioxins are chemically inert, water insoluble, stored in body fat, excreted in breast milk, and metabolized very slowly if at all. It almost certainly has a high chronicity factor because its whole-body half-life in rats is 24 days.

Perhaps dioxins or DBF were at one time associated with all polychlorinated aromatic mixtures or with any product that uses 2,4,5-trichlorophenol (TCP) as a starting material for its preparation. The evidence favors its introduction as a contaminant during synthesis rather than a result of a breakdown of the major ingredients in the environment, although the latter cannot be absolutely ruled out. Another product that uses TCP as the starting material is hexachlorophene, an ingredient in pHisohex, a widely used antibacterial soap and shampoo, surgical scrub, cleanser for newborn babies, and topical medication for acne vulgaris. In the early 1970s there was considerable alarm about the widespread and growing uses of hexachlorophene because it is known to be absorbed through the intact human skin and because of reports of brain damage when very high levels were fed to rats.

On July 10, 1976, an explosion occurred in a factory making TCP in Lombardy, Italy. The TCP was intended for export to a Swiss firm for use in the manufacture of hexachlorophene. A toxic cloud was released that contained very high concentrations of dioxins. The cloud settled over a large portion of the neighboring town of Seveso to create one of the largest problems in environmental decontamination ever faced by man. The total population of Seveso was evacuated. It was at first thought that at least they could serve as a base for some definitive epidemiological studies on the human health effects of dioxins. Unfortunately, this idea occurred too late for some critical baseline measurements, such as the collection of fat samples for the measurement of dioxin levels, so there was no way to relate actual exposure to any adverse health effects observed.

In exposed human populations that have been studied to date, there is no evidence to link dioxins to cancer or birth defects. Whether this is because the exposures were not sufficiently intense or whether it is because humans, like hamsters, are a very resistant species is not known. As already indicated, what is seen instead in exposed humans is mild liver damage, a loss of appetite, some nerve damage, chloroacne, and loss of libido, but there are those who believe that the present level of environmental exposure should not be a concern. Subtle effects on the immune system have been described, but their significance is presently unknown. In some species dioxins bind to receptors in the cytoplasm of cells. The dioxin-receptor complex then has access to the cell nucleus, and it can attach to DNA to turn on or off a complex series of biochemical switches.

Since 1971–1972, 1.4 million gallons of dioxin-containing Agent Orange have been stored in rusting steel drums on Johnson Island, a remote Pacific atoll, awaiting a plan for its safe disposal. Only in recent times has a course of action been planned. Specially built incinerator ships with furnaces capable of operating at 2400°C are intended to burn it and other toxic wastes at sea. This solution is somewhat ironic because dioxins have been identified in the fly ash from incinerators used to burn chlorine-containing trash at lower temperatures. The dioxins were apparently generated in the process. Although there is considerable interest in developing genetically engineered fungi and bacteria with appetites for persistent halogenated hydrocarbon pollutants and the hydrocarbons of oil spills, these are not yet ready for large scale use.

SUMMARY POINTS

No laws regulate the purity of most commercial chemicals; they are made in various purities. Technical grades are the cheapest because they contain the most impurities.

Some commercial chemical products are made as complex mixtures with dozens of the "desired" ingredients, but this compounds the problem of identifying impurities.

At various times a number of such chlorinated aromatic mixtures have been in widespread use in the U.S.; the best known are the PCB.

Because of their inertness and persistence in the environment, PCB concentrated in natural food webs, as did DDT and methyl mercury.

Many of these products appear to have been contaminated to varying degrees with highly toxic dioxins and DBF, which caused widespread losses of livestock.

The dioxins are highly toxic, teratogenic, and carcinogenic in some species, but it is possible that humans are among the more resistant to their effects.

BIBLIOGRAPHY

Coulston F Pocchiari F: Accidental Exposure to Dioxins. New York, Academic Press, 1983.

DiCarlo FJ, Seifter J, DeCarlo VJ: Assessment of the hazards of polybrominated biphenyls. Environ Health Perspect 23:351, 1978.

Kimbrough RD: Toxicity of chlorinated hydrocarbons and related compounds, a review including chlorinated dibenzodioxins and chlorinated dibenzofurans. Arch Pathol 94:125, 1972.

Kuratsume M, Shapiro RE: PCB Poisoning in Japan and Taiwan, Prog Clin Biol Res 137:1, 1984.

Tschirley FH: Dioxin. Sci Am 254:29, 1986.

Walsh J: Seveso: The questions persist where dioxin created a wasteland. Science 197:1064, 1977.

12

CHAPTER

HEAVY METALS

Bill D. Roebuck

Diseases associated with the mining and smelting of metallic ores or with the working of metals or their salts have been known for centuries. With growing industrialization these toxicologic problems became more widespread. New industrial, agricultural, and even medical uses for metals resulted in their dramatic redistribution within the environment and constituted new sources of exposure. Adsorption or surface binding of metals to certain types of soil, particularly clays, importantly influences their movement in the environment, such as the rate of leaching from toxic waste dumps. Like the halogenated hydrocarbon insecticides, some metals are persistent in the environment and are concentrated in natural food chains (see Chapter 10). As a result, poisonings began to occurr under bizarre circumstances such as the epidemic of methylmercury intoxication around Minimata Bay (see Chapter 9).

Heavy metals are natural constituents of the earth's crust, and trace amounts are always present in biologic materials. They are defined as electropositive elements having a density greater than five, but this arbitrary definition lumps together 40 metals of different properties and degrees of acute and chronic toxicity. Small amounts of zinc, iron, or copper are essential for biologic function. Molybdenum and titanium have little biologic activity. Mercury, lead, and cadmium are invariably toxic and a source of great environmental concern.

Some pairs of metals have mutually opposing biologic effects so that within limits the absolute level of environmental contamination may be less important than the relative levels of one metal to its antagonist. For example, in some animal species cadmium causes damage to the testes, and that effect is antagonized by zinc. Thus, over a wide range of concentrations the cadmium-zinc ratio is highly critical. In other cases the toxic metal has properties in common with an essential nutrient and poisoning results when the body attempts to use the toxic metal in normal processes. Calcium and lead constitute such a pair, and lead can be incorporated into bone and stored in the skeleton. In general, organisms have evolved elaborate mechanisms to assure adequate absorption of such essential metals as zinc and calcium, but toxic metals with similar properties may usurp these mechanisms and be absorbed in poisonous amounts. Finally, some essential elements, such as copper, when absorbed in excessive amounts can produce toxicity (see Chapter 5).

The major routes of exposure to metals are inhalation and ingestion. Inhalation is most common in the workplace. Mining, smelting, welding, and grinding are operations likely to produce finely divided dusts or fumes that can be inhaled. Attempts to protect workers against repeated or continuous exposures over their lifetime resulted in the setting of threshold limit values (see Chapter 4) and other standards. The combustion of leaded motor fuels or storage batteries results in atmospheric lead contamination, and its distribution throughout the environment. It is almost impossible to create a lead-free population of animals or humans for experimental study. Metals are invariably present in food and water, but these are rarely involved in poisonings in developed countries because of the variety of foods consumed and their origins from widely scattered geographic areas. When out of economic necessity reliance must be placed on a small number of local food items problems of poisonings are more likely to arise as, for example, in the case of Minimata Bay (see Chapter 9).

The chemical form of the element is important in determining its solubility in water and its eventual incorporation into biologic systems. The pure metallic element is usually insoluble in water and so unreactive as to resist incorporation. However, inorganic, ionic forms of the metal, as found in salts, will enter into biologic processes if they are sufficiently water soluble. Organometallic compounds such as methylmercury or tetraethyllead are particularly dangerous because of their enhanced ability to cross biologic membranes, including the skin and the so-called blood-brain barrier. The blood-brain barrier is thought to exist at the level of the endothelial cells, which line the blood vessels in the brain. Although it at least partially excludes charged, water-soluble ions and compounds, lipid-soluble organometallic compounds readily penetrate into the brain. This difference in distribution in the body accounts largely for the different forms of poisoning as produced by ionic and organic forms of the same metal. Similarly, differences in distribution as well as dose sometimes account for the different signs and symptoms in acute and chronic poisoning by the same form of a metal.

Ionic forms of some metals also have a high tendency to bind spontaneously and nonspecifically to biologic molecules, called ligands. Depending on the binding site this may result in toxicity or inactivation by storage in depots analogous to the storage of DDT in body fat (see Chapter 10). Table 12–1 illustrates several ligands, and shows that despite large differences in the binding strength or affinity, ligand binding is generally reversible. As noted above, there are distinct differences in the clinical pictures of acute

TABLE 12–1. TYPES OF CELLULAR LIGANDS THAT CAN INTERACT WITH METAL IONS AND THEIR BINDING STRENGTHS

	Dissociation Constant*	
Cellular Ligands	Methylmercury	Mercuric Mercury
Chloride (Cl^-)	5.4	6.7
Hydroxyl (HO^-)	9.5	10.3
Histidine	9.0	10.0
Cysteine†	16.0	14.0
Albumin†	22.0	13.0

*The dissociation constants are expressed as negative logarithms. The larger the number, the greater the tendency of the metallic ion to bind to that ligand. The free metallic ion exists in a state of equilibrium with the ligand bound form: ligand + metallic ion <----> ligand-metallic ion.

†These contain sulfhydryl groups (HS^-).

(After Clarkson TW: Recent advances in the toxicology of mercury with emphasis on the alkylmercurials. Crit Rev Toxicol 1:203, 1972.)

and chronic poisoning by some metals, such as mercury. It is likely that in acute poisonings, binding to ligands is simply a matter of which ligands are encountered first during absorption, which in turn depends on the route of administration. In chronic poisoning, however, the metal will gradually redistribute in the body to those ligands with the highest binding affinity. The ligands with the highest affinity are not necessarily those associated with the toxic effects, but if the redistribution results in signs of chronic toxicity, the signs may be very different from those of acute toxicity. Cells with high-affinity ligands associated with toxicity are called target cells because they accumulate metal and sustain cellular damage. Cells may also become targets because they have active transport mechanisms that can move the metal into the cell interior even against high concentration gradients.

A classic example of metal storage in the body is the deposition of lead in growing bones. Because of similar chemical properties, the body "confuses" lead with calcium and incorporates it into the bone matrix. Its presence is readily apparent in x-ray films of bones of children who live in lead-contaminated environments. In this form lead is not toxic, but it can exchange freely with calcium in the blood. If the body demands for calcium are increased, as in pregnancy, the demand is met temporarily by mobilizing calcium from bone. If the bone also contains lead, both lead and calcium may be released to the blood to result in signs of lead poisoning.

Metals undergo some limited biotransformation reactions. In the Minimata Bay incident microorganisms on the ocean floor were capable of converting ionic mercury to methylmercury, the most easily absorbed and most slowly excreted of the common mercury compounds. Tetraethyllead is slowly broken down in the human brain to inorganic lead. Just as inorganic lead penetrates into the brain poorly, when it is generated there by biotransformation, it penetrates back into blood only with great difficulty; thus, it is trapped in the brain. This storage form in a target organ is difficult to remove, and it results in long-lasting signs and symptoms. In animal cells and perhaps in human tissue, chromate ion in which the chromium has a valency of +6 undergoes biotransformation to chromic ion with a chromium valency of +3. Chromate rapidly crosses biologic membranes whereas chromic ion does not. Conversely, it is the chromic ion that reacts covalently with DNA, and that reaction is correlated strongly with chromate's ability to cause cancer in animals. Thus, it is thought that chromate penetrates into cells where it is biotransformed to chromic ion, which can then react with DNA.

In addition to nonspecific binding to expendable ligands, both plants and animals have small proteins and peptides that bind certain metals tenaciously and protect them against the expression of toxicity. These peptides and proteins contain sulfur, which has a high affinity for mercury and other metals (Table 12–1). Glutathione and metallothionine are examples of such metal-binding ligands (Figure 12–1). Exposure to certain metals leads to increases in the production and cellular concentrations of the ligands. Thus, like cytochrome P450, the sulfur ligands can be induced (see Chapter 7) to afford a greater degree of protection in response to the increased exposure.

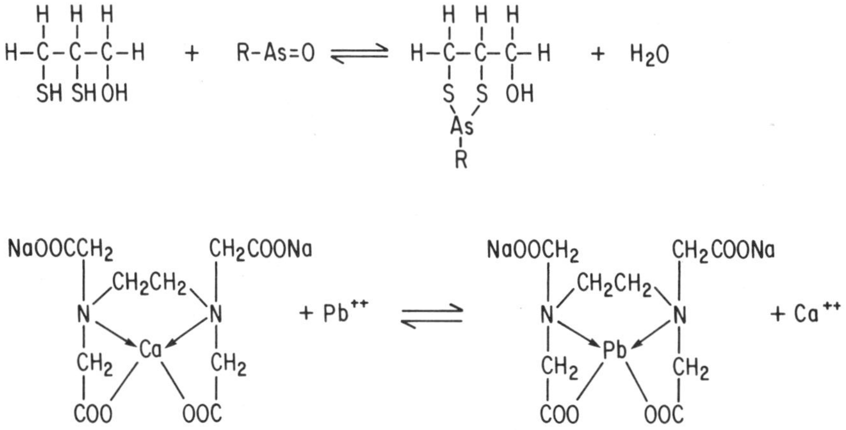

FIGURE 12–1. Therapeutically useful chelating agents. The agent at the top is dimercaprol or BAL. It is useful in arsenic, mercury, or gold poisonings. The agent at the bottom is edetate calcium disodium or $CaNa_2EDTA$. It is the agent of choice in lead poisoning.

Metal poisoning is extremely complicated because of the numerous ligands that can be involved. Almost all proteins contain sulfhydryl (-SH) groups, and all enzymes are proteins. At least 40 common intracellular enzymes can be inhibited by mercury, so the problem is not in finding what enzymes are inhibited, but in sorting out from among those known to be inhibited the ones that are the most critical for cellular functions. This phenomenon accounts in part for the plethora of toxic effects induced by metals. In addition to acute and chronic toxicity, certain metals are known to be either teratogenic or carcinogenic or both, and a number of them can induce allergic reactions as well.

The diagnosis, treatment, and prevention of metal poisonings will continue to be an important occupational and environmental concern. Rarely are specific signs or symptoms diagnostic of intoxication by a particular metal. Often the same signs and symptoms are also seen in other common medical problems. The chances for a correct diagnosis are much improved by a careful search of the patient's history for evidence of exposure to certain metals. If several persons with similar signs and symptoms are employed in the same activity within an industry, the diagnosis is strongly reinforced. Finally, the identification of high levels of toxic metals in the blood or urine of these persons makes the diagnosis a virtual certainty.

Once the diagnosis has been made, definitive treatment may be initiated. Acute poisonings are usually a medical emergency and immediate intervention is critical to alleviate pain or other discomfort and to ensure life. Afterward, removal from further exposure is essential. In chronic occupational exposures the victim must be reassigned to other activities, and this may be enough for spontaneous resolution of the illness. If not, medical treatment to reduce the body burden of a particular metal may be indicated. This may be accomplished by the administration of drugs known as chelating agents. "Chela" is the Greek word for claw. The "claw or claws" of chelating agents act like natural ligands to bind the metal tightly. The metal chelate is much more water-soluble than the metal ion itself so its urinary excretion is greatly facilitated. As shown in Figure 12–1, chelating agents often have multiple binding sites of more than one chemical bond with the metal to increase binding capacity and improve the stability of the chelate. This allows the drug to "steal" the metal away from its natural ligand. Although chelation therapy sounds simple in principle, it is complicated in practice, and prolonged and repeated courses of therapy may be needed.

Because of the difficulties in treatment, the prevention of heavy metal poisoning should be the ultimate goal. The key to prevention is education about the types of risks and durations of exposure, and provisions for avoiding or reducing exposure. If the sources of exposure can be identified, important steps can be taken. For example, the recognition that lead poisoning is an important health risk to growing children, and that an important source of exposure was lead pigments in paint resulted in a ban on such products. The removal of old, deteriorating lead paint surfaces was another measure. Nonleaded gasoline further reduces exposure, particularly in urban environments. Adherence to the TLV, monitoring of the air, adquate ventilation, use of masks or respirators, and the containment of vapors, fumes, and dusts all can be helpful measures in the workplace. For individuals at particular risk the monitoring of metal concentrations in blood or urine may be desirable.

Lead is the most common cause of human metal poisonings, and the most intensively studied. The literature on lead toxicity is enormous. Important sources of exposure are mining and smelting of lead ores or other ores contaminated with lead salts. The manufacture of storage batteries may involve exposure to lead. Lead water pipes like lead-based paints were once common, and may still be enountered in older buildings. Some glazes for pottery may contain lead pigments, which can leach into food or beverages. "Moonshine" or illicit whiskey may be contaminated with lead because old automobile radiators with lead solder or braze are sometimes used as the condenser part of the still.

Various forms of lead can be absorbed from the lungs, the skin, and the gastrointestinal tract. Significant absorption from the lungs requires penetration of the dust or fumes into deep lung structures (see Chapter 6), and a certain degree of water-solubility of the deposited material. In some cases, however, up to 40% of the deposited material may be absorbed with surprising rapidity. Organic forms of lead are absorbed much more rapidly and efficiently, but even these are absorbed only to a limited extent through the skin. The absorption of lead from the gastrointestinal tract is a complex process that depends on the type of lead compound consumed, the physiological state of the individual, and the type of food also present. In accordance with the general rule, lipid-soluble compounds are better absorbed, but some degree of water-solubility is also critical. Undissolved lead is absorbed to only a minor degree. The physical size of the particles is also important because small particles dissolve more readily than large particles.

The absorption of lead is also profoundly influenced by nutritional status with respect to calcium balance. When calcium demand is high, lead is more readily absorbed from the gastrointestinal tract because the mechanisms that have been mobilized to increase calcium absorption do not readily distinguish between calcium and lead. The type of food present in the gastrointestinal tract can also influence the extent of lead absorption. Fatty foods tend to enhance lead absorption although it is not entirely clear why they do so. Other materials in the gastrointestinal tract may bind or physically adsorb lead to decrease absorption. On average in humans, only 5 to 10% of the ingested lead is actually absorbed, and the rest is excreted with the feces.

After absorption, lead is distributed within the body in complex patterns. Some lead can be found in the liver, kidneys, and red blood cells, but more than 90% is rapidly incorporated into bone, apparently because of its chemical similarity to calcium. There it can be stored for long periods of time without toxicologic consequences. Unfortunately this deposit is in direct equilibrium with the concentration in blood and via the blood, indirectly with that in all other organs; when the body needs calcium from bone, some lead will also be mobilized and perhaps result in lead poisoning "from within." Thus, it can be seen that measurement of blood lead levels is an index of exposure to lead, but not an index of the total body burden, which is largely stored in bone.

Inorganic lead does not undergo extensive biotransformation in the body, as do organic lead compounds like tetraethyllead, nor does it undergo changes in valency, like the chromium ions above. In the human adult about 75% of the total lead excreted is found in the urine, and the remainder is excreted via the bile and feces. Traces are lost in sweat and from the natural flaking off of dead skin cells. In infants relatively more is lost in the feces and less in the urine. Organic lead is primarily excreted via the bile instead of the urine.

The major toxicologic effects of inorganic lead are exerted on the nervous system, the kidneys, and red blood cells from the bone marrow. The effects of lead on the brain fall into a group of disorders called encephalopathies, but inorganic lead may also affect nerves throughout the body (peripheral nervous system). Inorganic lead poisoning is charactrized by dullness of the senses, restlessness, tremors, and weakness in the extremities to the point where walking becomes difficult. There is great current interest in the effects of lead on learning ability in children. In contrast, organic lead primarily affects the brain with modifications in psychological behavior, excitement, delusions, and hallucinations.

Inorganic lead also damages the kidneys. One of the first effects is an inhibition of normal reabsorption of important nutrients from the fluid that is destined for processing into urine. Thus, these needed nutrients are lost in the urine. With much longer periods of exposure fibrous tissue is deposited in the kidney, resulting in a disruption of normal function, scarring of the kidney blood vessels, and shrinking of the glomeruli that initially filter the blood. The damage can progress to the point of kidney failure and death.

In the bone marrow lead affects the formation of hemoglobin and red blood cells so that one early sign of lead poisoning is anemia. One of the tests for lead poisoning is to look for the accumulation of intermediates in the hemoglobin biosynthetic pathway in urine or blood. The accumulation of these intermediates results in a disease called porphyria.

In 1946, in postwar Japan a physician returning to his private practice in Toyama Prefecture began seeing patients again with a painful affliction that was known locally as "itai-itai byo," which translates literally as the "ouch-ouch disease." The disease was more common in women than men. Most of the patients had small frames and grotesquely deformed spines. In some ways the disease was like osteomalacia, which may occur during pregnancy and is characterized by a softening of bone, including the spine, because of deficiencies of calcium, phosphorus, and vitamin D. Itai-itai, however, occurred in women past the childbearing age and some men.

A number of mines and refineries in the area produced copper, lead, zinc, and cadmium. Because arable land in Japan is scarce, every possible site is exploited for agriculture, particularly rice. The fields and paddies were sometimes irrigated with river water polluted by the tailings from mining or smelting operations. Crops could also be polluted by airborne metal emissions. In some areas chemical analyses of the rice showed cadmium at ten times the levels in unpolluted areas, and rice constituted a staple in the diet. There was also a high population density even in the relatively rural areas where the disease occurred so that a large number of people could have been exposed.

Exposure to cadmium and chronic cadmium poisonings had long been known to be associated with a number of industries. In acute poisonings the lungs and the kidneys were the target organs, and in chronic exposures the kidneys were primarily affected. Bone disease was not a prominent feature in industrially exposed workers. Thus, there was considerable controversy when the theory was first proposed that cadmium was the key contributing factor to itai-itai disease. An enormous amount of research and detailed epidemiologic studies finally convinced the authorities to designate it as a disease caused by cadmium in 1968. The number of patients officially designated as having the disease was 129, and no new cases have been reported since 1975. Most authorities believe that the true number of cases must have been substantially higher, but it may never be known with any certainty.

Nor, is it known with certainty why the signs of the disease in Japan were so different from those in industrially exposed workers. One theory is that the ingested cadmium damaged the lining of the intestines so that the absorption of important nutrients such as vitamin D, calcium, and phosphate was inadequate. Concomitantly, the damage to the kidneys, which itai-itai patients also exhibited, prevented the reabsorption of phosphate, calcium, and amino acids so that, as in lead poisoning, they were lost in the urine. Finally, one could add the additional complication of malnutrition because the victims were poor and subsisted mostly on rice. Because there are no new cases, itai-itai disease can no longer be studied in humans and no satisfactory animal model of it has ever been found. Therefore, it is unlikely that the hypothesis will ever be tested.

SUMMARY POINTS

Heavy metals are electropositive elements with a density greater than five; some are almost biologically inert, some are essential, and some appear to have only toxic effects.

Cadmium and lead have essential "twins," namely zinc and calcium, respectively, that can antagonize some of their toxic effects.

Organometallics (methylmercury and tetraethyllead) are dangerous because they penetrate the blood-brain barrier and the metal may be trapped in the brain.

Ionic forms of metals bind to a variety of natural ligands and eventually redistribute in the body to those that bind the most tenaciously, e.g., lead accumulates in bone.

Except for lead, metals show different syndromes in acute and chronic poisoning; chronic poisoning is difficult to treat because the metal is on its tightest binding sites.

Poisonings are treated with chelating agents that compete with natural ligands for the metal and bind it specifically as a water-soluble chelate for excretion in the urine.

BIBLIOGRAPHY

Clarkson TW: Recent advances in the toxicology of mercury with emphasis on the alkylmercurials. Crit Rev Toxicol 1:203, 1972.

Gosselin RE, Smith RP, Hodge HC: Clinical Toxicology of Commercial Products, 5th ed. Baltimore, Williams & Wilkins, 1984.

Hamilton A, Hardy HL: Industrial Toxicology, 3rd ed. Acton, MA, Publishing Sciences Group, 1974.

Hammond PB, Beliles RP: Metals. In Doull J, Klassen CD, Amdor MO (eds): Cassarett and Doull's Toxicology, The Basic Science of Poisons, 3rd ed. New York, Macmillan 1980.

Nelson N: Study group on mercury hazards. Environ Res 4:1, 1971.

Weeden RP: Poison in the Pot: The Legacy of Lead. Carbondale, IL, Southern Illinois University Press, 1984.

Wetterhahn K: Metabolism of the carcinogen chromate by cellular constituents. Struct Bond 54:93, 1983.

13

CHAPTER

THE SULFUR CYCLE AND REDUCING-TYPE AIR POLLUTION

Many chemical elements undergo cyclic redistribution in the environment from the atmosphere to land and oceans and back again. Only two of these have been selected for discussion here, sulfur in this chapter and nitrogen in the following one. They were selected, first, because they are probably the most completely understood although surprising gaps still exist in our knowledge. Second, because human activity is resulting in important disturbances in these cycles, and third, because various chemical forms of sulfur and nitrogen are known to have adverse human health effects and damaging effects on the environment.

CHEMISTRY, PRODUCTION AND USES OF SULFUR

The discovery of sulfur, which is said to be the ninth most abundant element in the universe, and the finding of uses for it antedate recorded history. It is referred to repeatedly in The Bible as brimstone, the stone that burns. Since very ancient times the smoke and fumes from burning sulfur have been used in religious ceremonies to repel evil spirits. It was also used as a fumigant, bleach, and insect repellent.

Sulfur is mined by the Frasch process, in which a special pipe with three concentric layers is lowered through a previously bored hole into a sulfur deposit. Superheated steam is then pumped down the outermost pipe to melt the sulfur, which is insoluble in water. Compressed air is forced down the innermost pipe causing the molten sulfur to rise through the middle pipe to ground level, where it is poured into forms to allow the water to evaporate and the sulfur to harden and set. Layer is poured on layer in this fashion resulting in a structure known as a vat of sulfur. This was the major source of sulfur through the 1950s. Although it is still used today, it has been surpassed by sulfur recovery from petroleum and natural gas. This recovery process begins at the wellhead with the stripping out of hydrogen sulfide.

About 90% of the sulfur produced is used to make sulfuric acid (H_2SO_4) and the lion's share of that is used in turn to make fertilizer. The production of sulfuric acid by a country is an indicator of economic development and technological growth. In many countries the volume of sulfuric acid produced parallels closely the gross national product. Sulfur, limestone, salt, coal, and oil constitute the five basic raw materials of the chemical industry. In fertilizer production equal parts of sulfuric acid and crushed, insoluble phosphate rock are mixed together to yield soluble superphosphate fertilizer. Gypsum (calcium sulfate) is a byproduct.

Elemental sulfur, designated as S°, can be both oxidized and reduced (Table 13–1). Examples of compounds are known in which the valency of the sulfur varies from −2 to +6, which is an unusually large number of stable valency states. All of the chemicals listed in Table 13–1 are believed to play roles in the natural sulfur cycle.

THE SULFUR CYCLE

The natural sulfur cycle is illustrated in Figure 13–1. Both volcanoes and natural gas emissions contribute to atmospheric hydrogen sulfide. If the volcano is actually erupting and the hydrogen sulfide is exposed to air, some or all of it will be burned to yield sulfur dioxide (SO_2). Indeed, volcanoes appear to be the dominant source of nonanthropogenic SO_2 (and eventually sulfuric acid, as discussed below) in the stratosphere. The Mexican volcano, El Chicon, erupted twice (on March 20 and April 4, 1982), which made it the largest volcanic eruption in recorded history since Krakatoa in 1884. Measurements in the stratosphere before

TABLE 13–1. THE STABLE OXIDATION STATES OF SULFUR

Sulfur Formula	Name	Valency
H_2S	Hydrogen sulfide	−2
S°	Sulfur	0
$Na_2S_2O_3$	Sodium thiosulfate	+2
SO_2	Sulfur dioxide	+4
H_2SO_4	Sulfuric acid	+6

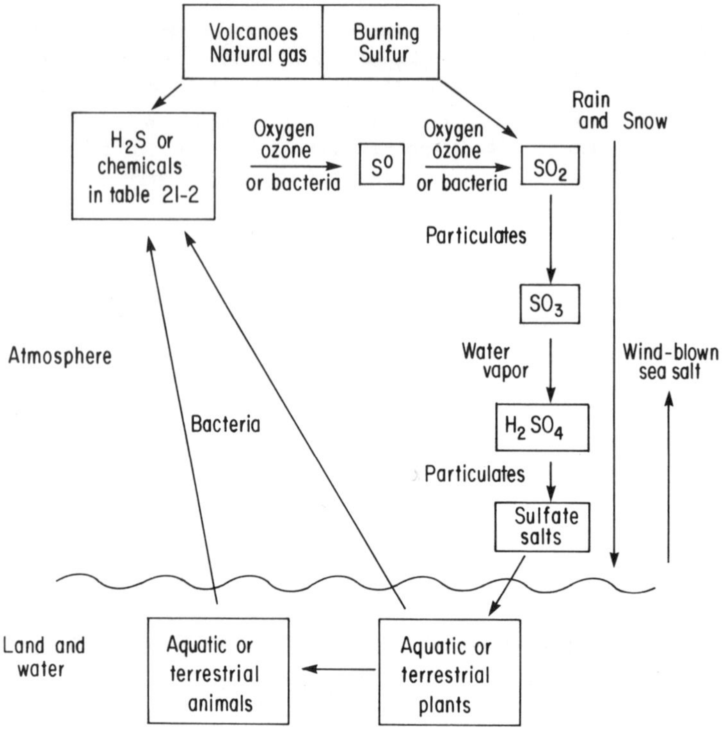

Figure 13–1. The natural sulfur cycle. (See text for details.)

and after the eruptions of El Chicon showed substantial increases in the atmospheric concentration of various sulfur oxides (SO_x).

Other sources of naturally burning sulfur also contribute to atmospheric SO_x. In the Cape Bathurst area of the Northwest Territory of Canada a 30-km stretch of sea cliffs along the Arctic Ocean, known locally as the smoking hills, consist of spontaneously burning deposits of bituminous shale. As the cliffs erode and slide into the sea, fresh deposits of coal are exposed for burning. These deposits have been burning for hundreds, and perhaps even thousands of years. The smoke is extremely dense, and it contains high concentrations of SO_x and sulfur acids. The intense acid fumigation has profoundly influenced the local tundra. In other parts of the Northwest Territory the ponds are typically alkaline with pHs of 8.0 or higher, but in this area some ponds had a pH of 2.0, which is extremely acid. This area may represent in microcosm an extreme of the now widespread problem of acid rain (see below).

These natural sources of atmospheric SO_x tend to be widely distributed over the earth, and for obvious reasons, they are generally in sparcely populated areas. As illustrated in Figure 13–1 atmospheric SO_2 is oxidized in a series of steps to H_2SO_4. The H_2SO_4 may be formed on the surface of atmospheric particulate matter, which acts catalytically to synthesize it from SO_x and water vapor, or it may encounter airborne particles after it has been formed. In contact with the particle, the H_2SO_4 forms sulfate salts, and both the salts and the acid are carried to the earth's surface by precipitation, although dry deposition occurs to a limited extent. The sulfate ion is taken up by plants, and the sulfur is incorporated into plant proteins. When the plants die, bacterial action converts the sulfur into volatile forms such as H_2S for a return to the atmosphere. Alternatively, the plants may be consumed by animals, the proteins broken down, and the sulfur reincorporated into animal proteins. Part of the sulfur is recycled from animal to animal and part is returned again to the atmosphere by the actions of putrifying bacteria. In an atmospheric process, H_2S is oxidized spontaneously to SO_2 by oxygen or ozone or with the help of airborne bacteria.

If the concept of a natural sulfur cycle has validity, it is certainly necessary to postulate the formation of a volatile chemical form of sulfur on earth or in the oceans to return the sulfur to the atmosphere. The National Center for Atmospheric Research (NCAR) in Boulder, Colorado, has formulated a model of the natural sulfur cycle in which H_2S is importantly involved (Kellogg et al., 1972). When something is known about the size of the various compartments in the sulfur cycle (Fig. 13–1), predictions can be made about what the atmospheric concentration of H_2S would have to be in order to sustain a balance. One such estimate (Lovelock et al., 1972) is that an average atmospheric concentration of 20 ppb (parts per billion) of H_2S would satisfy the mass transfer needs of the NCAR model.

There were numerous technical difficulties associated with the measurement of very low levels of atmospheric H_2S. When these were finally solved by the NCAR, the results of measurements in the relatively unpolluted air over Boulder showed that H_2S was present in the ppt (parts per trillion) level instead of ppb. Thus, its concentration was as much as 1000 times too low to satisfy the needs of their own model. This finding prompted numerous searches for alternative volative forms of sulfur that might play important roles in the transfer of sulfur from the earth to the atmosphere.

One interesting suggestion (Lovelock et al., 1972) is dimethylsulfide or DMS (Table 13–2). Many living things generate DMS, and prominent among these are certain marine algae. The measured concentration of DMS in sea water would be sufficient to sustain an atmospheric concentration in the low ppb range, which is much closer to the NCAR predictions than is H_2S. Soil samples

TABLE 13–2. VOLATILE FORMS OF SULFUR POSTULATED TO PLAY A ROLE IN THE NATURAL SULFUR CYCLE BY MOVING SULFUR FROM THE OCEANS OR EARTH TO THE ATMOSPHERE

Formula	Name
H_2S	Hydrogen sulfide
CH_3-S-CH_3	Dimethyl sulfide (DMS)
COS	Carbonyl sulfide
CS_2	Carbon disulfide
CH_3SH	Methyl mercaptan
CH_3-S-S-CH_3	Dimethyl disulfide

collected from widely dispersed geographic locations within the U.S. all spontaneously emitted DMS. Presumably these emissions were a result of bacterial action on natural sulfur compounds found in the soil. Dr. Lovelock and his associates believe that if these rates measured in U.S. soil samples are representative for the rest of the earth, DMS could sustain the balance in the natural sulfur cycle. Others have since shown that both DMS and H_2S are spontaneously emitted from salt marshes. There are at least theoretical pathways by which DMS could be oxidized in the atmosphere to SO_2 so that it is possible to reconcile the Lovelock hypothesis with the NCAR model.

The initial step in the hypothetical atmospheric oxidation of DMS would be a conversion to dimethyl sulfoxide (DMSO). This interesting chemical has received a great deal of attention within the past twenty years as a miracle panacea or cure-all drug. More critical study has refuted most of the claims, but it still has a folk medicine reputation for usefulness in arthritis. Perhaps the best that can be said for DMSO is that at least up till now it is not recognized to cause any serious harm. It is practically nontoxic when given acutely to laboratory animals.

As shown in Table 13–2 at least four other candidates have been proposed for the volatile link in the sulfur cycle. Of these, carbonyl sulfide and carbon disulfide have also been shown to emanate spontaneously from salt marshes. Thus, the current problem is not so much to identify potential volatile links in the sulfur cycle as it is to show which of these is quantitatively the most important. The most recent evidence, however, favors carbonyl sulfide as the dominant nonvolcanic stratospheric source of SO_2 and H_2SO_4 (Hofman, 1990).

It is surprising that the answers to such questions are not known with greater certainty since these processes are responsible for the transfer of over a half-billion tons of sulfur each year (Kellogg et al., 1972). The bulk of that ascending into the atmosphere is believed to result from the postulated volatile link. Smaller, but still substantial, amounts arise from windblown sea spray (Fig. 13–1). Nearly everyone at some time or another has experienced the smell of an ocean even before it could be seen. The composition of the ambient atmosphere in costal areas where there is an active surf resembles that of dilute sea water.

Sulfur Dioxide

The human activity that produces the largest intrusion on the natural sulfur cycle or about 70% of all the anthropomorphic sulfur is the burning of fossil fuels, especially coal. This activity, of course, occurs in the areas of densest population, which are also the areas of the most intense industrial activity. Such areas include the American midwest and the northeast corridor, central England, the Ruhr Valley in Germany, Poland, Czechoslovakia, and what was formerly southern East Germany. One of the things found behind the lifted iron curtain and the demolished Berlin wall was one of the most polluted areas on our planet, and a population poorly equipped technologically and economically to deal with pollution problems. Second to carbon monoxide (CO), SO_2 is the single most prevalent air pollutant in metropolitan atmospheres.

This anthropomorphic contribution, which is appoaching the size of the naturally ascending pool of atmospheric sulfur, differs from it in that almost 95% is already in the form of SO_2. The NCAR estimates that by the year 2000 man's contribution to the total atmospheric sulfur in the northern hemisphere may exceed that of nature. Because the known reserves of low sulfur coal have been largely consumed already, some of the hopes for the future are that an economically feasible method will be found for removing sulfur from coal before it is burned, or for more efficient removal of SO_x from flue gases after burning, i.e., better scrubbers. Indeed, one new wet scrubber sprays the smoke with water and limestone to make gypsum for wallboard. Two new interesting coal technologies that are being developed are pressurized fluid bed combustion capable of reducing SO_2 emissions by almost 90% with a higher thermal efficiency, and mild coal gasification to convert subbituminous coal into a better solid fuel with liquid fuels as byproducts.

Hydrogen Sulfide

Of the chemical forms of sulfur believed to participate in the sulfur cycle, H_2S, SO_2 and H_2SO_4 are the most dangerous acutely. However, H_2S has little or no tendency to produce chronic or cumulative effects. Its high acute toxicity and its low chronicity index are properties like those for hydrogen cyanide and its salts (see Chapter 4). Indeed, these two chemicals are believed to share a common mechanism for their acute toxicity (see discussion below).

Air pollution by H_2S tends to be a localized problem because most emission is at point sources, such as paper mills, tanneries, waste disposal ponds, petroleum refineries, sewage treatment plants, sources of geothermal energy and coke ovens. Some deposits of natural gas contain as much as 90% H_2S. Because it is not a widespread urban pollutant, the EPA has not set a national air quality standard. Several states and other countries, however, have set such standards. Even at very low concentrations (0.1 to 0.2 ppm) H_2S has an extremely offensive odor, like that of rotten eggs. The basis for ambient air quality standards is usually odor, and the standards are commonly 0.1 ppm or less. The Occupational Health and Safety Agency (OSHA) and the American Conference of Governmental Industrial Hygienists are responsible for setting standards for the workplace. For occupational exposures over a lifetime, the TLV is 10 ppm, and it was set on the basis of eye irritation. At about 1000 ppm H_2S can cause death within a few minutes.

Rotorua is a city of about 40,000 people situated on the North Island of New Zealand in an area of extensive natural geothermal activity. The hot springs and numerous holes or vents in parts of the city bring to the surface steam and gases including H_2S. The odor of the latter is present everywhere in the city, as it has been since it was first settled by the Polynesians thousands of years ago. Although a few accidental deaths have occurred as a result of localized buildups, no adverse, long-term, chronic effects have been ascribed to exposures over the lifetimes of many generations. The geothermal energy is extensively exploited by the current residents for cooking and home heating. The waters are used in spas for their warmth and alleged healthful effects. Indeed, Rotorua is a tourist attraction. Thus, tangible economic benefits compensate for the odor nuisance. One wonders if the residents would remain so openly enthusiastic about their community in the absence of these economic advantages. Obviously, public attitudes toward chemicals may be strongly influenced by factors other than concerns about their health effects.

An interesting example of the adaptation of a primitive life form to high concentrations of H_2S is exemplified in the so-called hydrothermal vent, tube worm. These creatures were discovered recently in colonies in the Galapagos Rift west of Ecuador at a depth of 1.5 miles. They fluorish in the areas around active fumaroles, which are venting volcanic gases. Ordinarily, sea water at such depths is near freezing, but near such vents the temperature may be as high as 10°C. The gases contain high concentrations of H_2S, and the average concentration found in the blood of the worms was 0.54 mM. From laboratory animal studies it is known that the lethal concentration of H_2S in mammalian blood is on the order of 0.05 to 0.1 mM. Thus, these worms are living happily with 5 to 10 times the blood concentration that would be lethal to more conventional animals.

It is generally believed that both H_2S and HCN exert their rapidly lethal effects by inhibiting a critical enzyme in aerobic animals, cytochrome oxidase. Cytochrome oxidase is necessary for the utilization of molecular oxygen at the cellular level. In a series of elegant experiments it was shown that vent worm cytochrome oxidase was just as sensitive to inhibition by H_2S as the mammalian enzyme. Because the effects of H_2S in whole tissues was the same as in homogenized tissues, it was concluded that there were no natural barriers in tube worms that kept the inhibitor away from the enzyme.

Eventually, it was discovered that primitive tube worm hemoglobin, unlike normal mammalian hemoglobins, binds sulfide with a different ligand from that which binds oxygen in either worms or animals. The bound H_2S is then transported to a unique internal organ where symbiotic bacteria utilize the sulfur and metabolize it to nontoxic thiosulfate, thereby accounting for the resistance of the tube worm to its lethal effects. It has recently been suggested that thiosulfate may play a more important role than previously recognized in a cyle within the sulfur cycle that exists in anoxic marine sediments (Jorgensen, 1990). Of the remaining chemicals in Table 13–2, gaseous carbonyl sulfide also has a high acute toxicity in animals because it is converted to H_2S in the body.

Other Volatile Sulfur Compounds

The volatile DMS is considerably less toxic than either of the above. It has been approved by the FDA for food use, and in some countries it has the equivalent of GRAS status. In the rabbit it is biotransformed to DMSO as in the hypothetical scheme above for its atmospheric oxidation. Carbon disulfide is used in the viscose rayon industry, and it is known to be extremely dangerous. It has a high tendency to produce chronic neurologic and psychiatric illnesses. Fortunately, the environmental levels that have been measured are well below those associated with central and peripheral nerve damage in occupational exposures.

Methyl mercaptan (Table 13–2) is a relatively toxic gas with an extremely pungent odor. It is sometimes added in low concentrations to manufactured or natural gases used as fuels to serve as a warning of leaks. It has been hypothesized that methyl mercaptan may be generated in humans in some forms of liver disease and may be responsible for the associated coma. In some species it is partially biotransformed to DMS. Dimethyl disulfide (Table 13–2) is partly converted both to DMS and to methyl mercaptan in mice.

SULFUR AND PARTICULATES

In terms of both acute and chronic adverse health effects of urban air pollution in northeastern U.S. cities, suspicion has long been focused on SO_2. It is well known as a respiratory tract irritant, and its suffocating odor is detectable in the range of 3 to 5 ppm. It is dangerous even in short exposures at 200 to 500 ppm. In high concentrations it may produce spasm or edema of the glottis, halting airflow into the lungs and producing rapid suffocation. The irritant effects at lower concentrations may result in a more slowly developing lung edema (pulmonary edema or the adult respiratory distress syndrome).

In laboratory studies to evaluate the toxic effects of chemicals considerable pains are usually taken to ensure that the purest available form of the chemical is used. As indicated in Chapter 11 highly toxic impurities, even when present in vanishingly small amounts, can lead to gross misinterpretations of the results. In a one-year study, guinea pigs were exposed to pure SO_2 at three different concentrations for 22 hours per day, 7 days/week for 1 year. The highest level used in the study was 5.7 ppm. A sophisticated battery of observations and tests failed to reveal any adverse health effects that could be ascribed to the exposure.

A totally unexpected finding in the study was an increased survival rate for guinea pigs exposed to 5.7 ppm SO_2 when compared to the control group that was exposed only to air. The SO_2-exposed group also had a lower incidence and decreased severity of spontaneous diseases of the lungs and upper airways. Eventually, this result was ascribed to an antiseptic effect of SO_2 in that it destroyed airborne bacteria or viruses, or perhaps even such organisms within the airways, which might otherwise have given rise to fatal infections. It is likely that this effect of SO_2 is demonstrable only over a narrow range of concentrations. Certainly, it cannot be regarded as an unqualified endorsement of the health benefits to be obtained from breathing the gas at low concentrations.

Three well documented episodes of acute air pollution resulted in increases in the human death rate: in 1930 in the Meuse Valley of Belgium; in Donora, Pennsylvania, in 1948; and in London in 1952. These three incidents had many features in common. The necessary meteorologic condition is called an inversion. Normally, when the layers of the atmosphere closest to the earth become warmer than the upper layers, they rise and become mixed and diluted as the colder air descends. The opposite conditions prevail in an inversion; the layers closest to the earth are colder than the upper layers and do not rise. This creates a stagnant air mass as in a pot with a lid. Multiple sources of air pollution in the area then act to build up pollutant concentrations. In each of the above cases the inversion lasted for only 3 or 4 days, however, dangerous concentrations of pollutants in some areas can build up rapidly. In each case coal was the major fuel in the area. The pollution became known as reducing-type pollution, in which SO_2 is a prominent ingredient.

The total number of deaths in the Meuse Valley was 65, in Donora 20, and in London approximately 3200. Not all of these deaths, however, could be ascribed with absolute certainty to the pollution incidents. When the data were analyzed by statisticians, they were expressed in terms of so-called excess deaths, i.e., the number of deaths above and beyond the normal death rate for that time of year in that area. When plotted in this way (Fig. 13–2), it can be seen that there was a significant increase in death rate coinciding with the onset of pollution. The death rate then began to fall as the pollution abated. Surprisingly, however, a compensatory dip in the death rate occurred in the several days following the episode that was almost exactly equal in magnitude to the initial increase.

There were no pathologic findings among the victims that could unequivocally be ascribed to exposure to air pollutants. In gen-

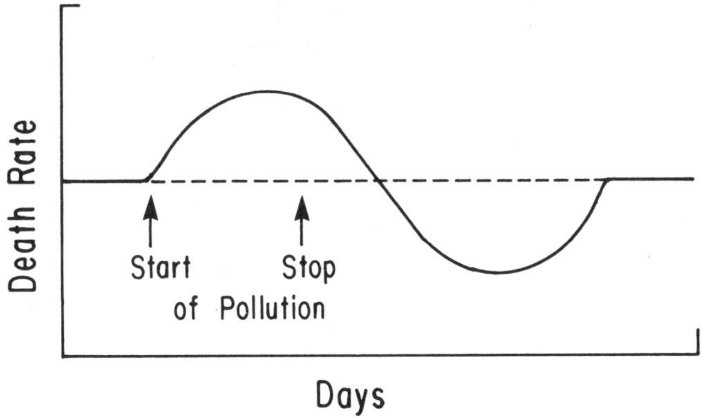

Figure 13–2. Hypothetical fluctuations in the natural death rate during acute episodes of reducing-type air pollution (See text for details.)

eral, the deaths tended to occur among the elderly and particularly those with pre-existing heart or lung disease. These findings strongly suggested that the deaths occurred among individuals who were already in poor health and perhaps close to death. The added stress of the pollution episodes may have carried them away a few days or weeks earlier than might otherwise have been the case.

The other significant finding was that the SO_2 levels in these incidents averaged 0.4 ppm. Even on the worst day in London, the average was only 1.3 ppm. These numbers stand in contrast with the laboratory study above in which 5.7 ppm of pure SO_2 was associated with increased survival rates in guinea pigs. Although the epidemiologic studies were based on acute episodes and the animal study was chronic, it still appeared that a laboratory study with pure SO_2 was not a good model for evaluating the adverse health effects of reducing-type urban air pollution.

In terms of lifetime adverse human health effects of reducing-type air pollution, rigorous and quantitative cause and effect relationships have been very difficult to establish. It is widely held to be contributory or aggravating to such diseases as the common cold, chronic bronchitis, chronic obstructive lung disease, emphysema, asthma, and cancer. All of these are also caused or aggravated by tobacco smoking. Clear chemical differences exist between tobacco smoke and urban air pollution, but both are such complex mixtures that they defy complete chemical description. In some studies in which the smoking variable had been controlled, air pollution accounted for small but measurable contributions to the incidences of the above diseases.

The startling discrepancy between laboratory studies with pure SO_2 and epidemiologic studies, in terms of the levels associated with adverse health effects, led gradually to the realization that the former may be almost irrelevant to the latter. Reducing-type air pollution seemed to be considerably more hazardous than comparable concentrations of pure SO_2, so searches were begun for more dangerous ingredients in air pollution or for ingredients that might potentiate the toxicity of SO_2. Urban air pollution consists of thousands of distinct chemicals, and the composition varies with geographic area, meteorologic conditions, season, and even time of day. All urban atmospheres, however, have one feature in common, namely, suspended particulate matter that is invariably emitted along with SO_2 when coal is burned. These are responsible for the smoke and soot. Particulates are also contributed from other sources as well, including rubber particles from tires, asbestos and other fibers from brake linings, and dust from construction sites.

It took a radical departure from the practice of using pure chemicals for laboratory studies to unravel this mystery of urban air pollution. It was only when scientists began working with combinations of SO_2 and aerosols (semistable suspensions of particulate material in air) that laboratory results began to agree more closely with the epidemiologic studies. A well known acute effect of SO_2 exposure in humans and laboratory animals is an increase in airway flow resistance resulting from reflex bronchoconstriction. Humans and guinea pigs appear to be particularly sensitive to that effect. The details of how such measurements are made are beyond the scope of this discussion, but the validity of the method has been generally accepted for many years.

Perhaps the first demonstration of a synergistic effect between particulates and SO_2 was made by Dr. Mary Amdur and Dr. Morton Corn in 1963. They were acting on an earlier suggestion that zinc ammonium sulfate particles were in part responsible for the irritant nature of the pollution in the Donora incident. This soluble salt had been identified from samples of particulate matter removed from an air conditioning unit filter that had been in operation at the time of the inversion. Unfortunately, there had been no actual measurements of airborne particulates during the Donara tragedy, but measurements during a subsequent less severe incident of local pollution allowed some estimates to be made about what the levels might have been.

Using these estimates, guinea pigs were exposed for one hour to aerosols of various particulates with and without SO_2. As shown in the top panel of Figure 13–3, the particulate aerosols alone turned out to be irritating and to produce significant increases in airway flow resistance. The three aerosols were adjusted such that each had about the same average particle size and the same airborne concentration of particles (about 1 mg/m^3). On this basis zinc ammonium sulfate was the most irritating, followed by zinc sulfate and ammonium sulfate. The results suggested that the irritant properties of the three aerosols were not specifically caused by zinc or ammonium or sulfate, but the chemical identity of the particles did in some way influence the irritancy. For example, H_2SO_4 mists of the same particle size are much more irritating than any of the aerosols in Figure 13–3.

In further work, a concentration of zinc ammonium sulfate aerosol was selected, which produced a measurable but submaximal effect. Various concentrations of pure SO_2 were then tested until a concentration was found that produced an effect identical to the zinc ammonium sulfate response (bottom panel, Fig. 13–3). When the combination of the particulate aerosol and SO_2 was tested, the effect was more than double that of either zinc ammonium sulfate or SO_2 alone even though the SO_2 in the mixture was lower than the concentration used alone. Doubling the concentration of the particulates alone did not result in an effect as

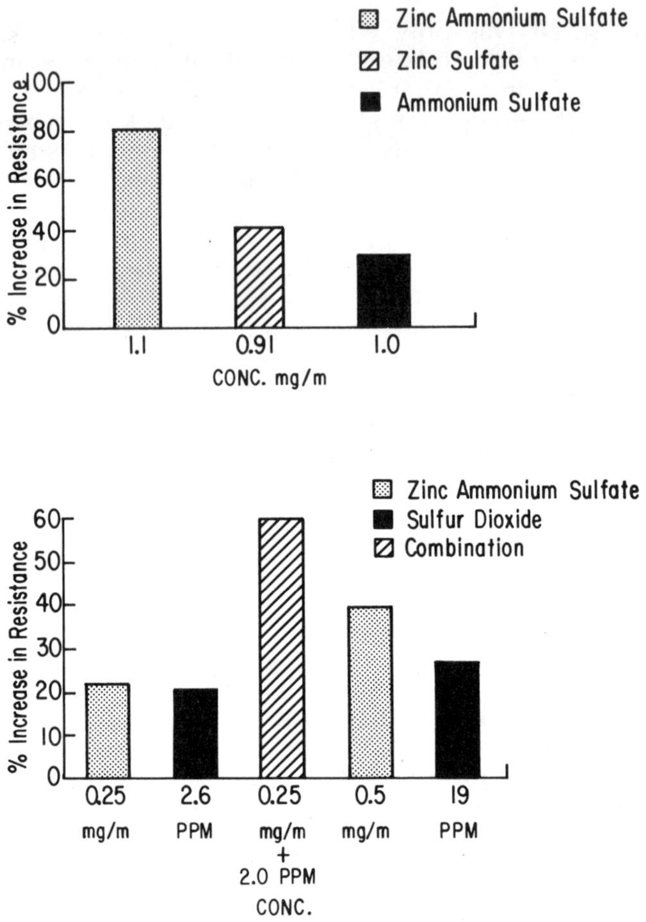

Figure 13–3. Top. Comparison of responses of the guinea pig airway to equal concentrations of zinc ammonium sulfate, zinc sulfate, and ammonium sulfate. Bottom. Responses to zinc ammonium sulfate and sulfur dioxide alone and in combination. (See text for additional details.) (From Amdur MO, Corn M: The irritant potency of zinc ammonium sulfate of different particle sizes. Am Indust Hyg Assoc J 24:326, 1963.)

great as the combination nor did increasing the concentration of SO_2 alone by more than sevenfold.

Within certain limits, particle size was also found to influence the irritant response. When the amount of chemical in the aerosols was held constant and the particle sizes were made smaller and smaller, airway flow resistance was found to increase. Thus, within the range of particle sizes studied, smaller particles were more hazardous than larger ones. At the same time, however, because the chemical concentration was held constant, the finer aerosols necessarily contained larger numbers of particles per unit volume than did coarser aerosols. When large and small particles were compared at equal numbers per unit volume instead of equal chemical concentrations, however, the greater response was elicited by the larger particles. The reasons for this latter confounding observation are not clear, but it was concluded that if the zinc ammonium sulfate particles at Donora had been in the range of 0.3 to 0.5 microns, they probably contributed to the irritant effect, whereas if they had been larger than 1.4 microns, they probably did not.

At least three possible reasons can be offered for the greater hazard of smaller particles at equal masses with larger particles. The first is probably that the smaller particles are carried into deeper and more vulnerable lung structures (see Chapter 6). Second, in combination with SO_2 some of the gas is probably adsorbed on the particle surface. Thus, the particles serve as carriers for SO_2 some of which might otherwise have been scrubbed out by the upper respiratory tract. The final reason has an analogy with the natural sulfur cycle (Fig. 13–1). When SO_2 adsorbed on particulate surfaces comes in contact with water vapor in the respiratory tract, the production of H_2SO_4 is catalyzed, and this acid is by far the most irritant form of sulfur. The finer the aerosol for a given mass, the larger the surface area that is available for this catalytic effect.

In laboratory studies it is common to assume that particulates in aerosols are perfect spheres. Almost certainly this is not true of urban particulates and therefore particle shape may be another variable in eliciting the response. In any event, ambient air quality standards are no longer defined in terms of the SO_2 level alone, but in terms of both the SO_2 and the total particulates, a change that was clearly in the public interest.

Woodsmoke

During the first OPEC-inspired oil shortage many people in northern rural areas in the U.S. returned to the wood stove as a more reliable and economic source of heat for their homes. The environmental and health impact of domestic woodburning became a new issue. One of the few studies conducted on the subject was led by Professor James Hornig of Dartmouth College. He selected a small nearby community where more than 75% of home winter heating was done by wood. This work identified and measured air concentrations of a number of known carcinogens in the family called polycyclic aromatic hydrocarbons. Even in this pristine New Hampshire environment, however, PAH could originate from nearby and remote coal combustion, oil refining and other industrial processes. The problem was to sort out the specific contribution of woodsmoke from these other sources. What was needed was an airborne marker specific for woodsmoke. Further work uncovered one substance, a sugar-related molecule called levoglucosan, which arises only from the incomplete combustion of wood. With this unique and quantitative marker for woodsmoke, its contribution to the total atmospheric PAH could be predicted even in the presence of the other sources.

Absorbed Sulfites

For many years the adverse health effects of SO_2 were considered only in terms of the airways and lungs, but that was somewhat myopic. It has long been known that exposed workers showed effects that must have been caused by systemic effects of the absorbed gas, such as a high urinary acidity. As in the natural sulfur cycle, it seemed likely that absorbed SO_2 would be hydrolyzed to sulfurous acid (H_2SO_3) and excreted in the urine. More recent studies with radioactively tagged SO_2 showed clearly that the inhaled gas is at least partially absorbed and distributed throughout the body. The H_2SO_3 in blood ionizes to hydrogen ion (H^+) and sulfite anion (SO_3^-). The latter can react reversibly with plasma proteins, which serve as temporary storage sites.

Exactly the same thing happens if one feeds or injects sodium sulfite (Na_2SO_3) instead of inhaling SO_2. Various sulfite salts are widely used GRAS food additives for their antimicrobial and their antioxidant properties to prevent the "browning" reaction. They are used extensively in making wines, and they may be present in substantial amounts in the final product. Sulfiting agents are also added to dehydrated fruits, vegetables, soups, fruit juices, and beer. Some drugs contain added sulfite as a preservative, and sulfites used as pesticides may result in residues on foods. It was once used widely in salad bars as a preservative, but that practice is actively discouraged today. Exposure is insidious because of the multiple and diverse sources, and the total intake is measured as sulfur dioxide equivalents. One national review group concluded recently that for most people the current levels and use patterns do not constitute a health hazard. A small fraction of the population, however, is known to be sulfite-sensitive and may experience reactions of unpredictable severity. These reactions probably have an allergic basis, but the sensitivity may also involve an inherited trait.

Acid Rain

The area of greatest current public concern in relation to the sulfur cycle is that of acid rain and its alleged association with alarming and widespread forest dieback, which is clearly evident across huge areas of North America and northern Europe. Acid rain (or snow) is defined as precipitation with a pH of less than 5.6. It is essentially a dilute mixture of H_2SO_4 and nitric (HNO_3) acids. The latter arises also from the burning of fossil fuels (see Chapter 14). In an extreme example, acid rain that fell in Scotland had a pH of 2.4, which is the acidic equivalent of vinegar. In contrast, melted ice cores from Greenland glaciers and ice sheets that had fallen as snow around 1800 had pH values ranging from 6.0 to 7.6. Analysis of such cores has shown that the sulfate concentration has tripled since the early 1900s and that the nitrate concentration (see Chapter 14) had doubled since about 1955. In addition to forest damage, acid rain is widely held to be responsible for changes at all levels in food webs in lakes and ponds. Without ice cores, it was difficult to establish the past acidity history of such open bodies of water until diatoms from sediment cores were examined. The past acidity can be estimated from the number and type of diatoms in sediment layers. In Adirondak lakes the acidity may have changed by one pH unit over hundreds to thousands of years, but some had changed by that much in the past 20 to 40 years because of acid rain.

Unfortunately, as is so often the case, the scientific evidence about acid rain has lagged woefully behind what might now be rapidly developing and potentially irreversible catastrophic events. Some of the earliest observations were made by scientists at Dartmouth College working at the Hubbard Brook Experimental Forest. It was noted that upland streams were showing an increase in acidity and a decrease in bicarbonate. In control streams bicarbonate was the dominant anion, but in the Hubbard Brook area it had been completely replaced by sulfate (Johnson et al., 1972). It is a complex issue and almost every aspect has contributed to the controversy. The problems in the northeastern U.S. and the Maritime Provinces of Canada have been blamed on emissions from coal-burning power plants in the midwestern U.S. Even the EPA has been criticized for possibly contributing to the problem. When some of these midwestern power plants were found to be in violation of emissions standards, one solution was to build taller smoke stacks. Some now are as much as 1000 feet in heighth. This "solution" put the emissions higher into the atmosphere where they can be carried further away, but the local problem was thereby converted into a regional, continental, and perhaps even global one. Chemical analyses of pollution aerosols in the northeast, however, suggest to some that although midwestern sources may dominate in an episodic manner and perhaps contribute regularly to the total amount, sources in the northeast area itself may be comparably important for the long-term. Often little or no correlation exists between changes in SO_x emissions in a given locality and changes in the acidity of surface waters.

Whether acid rain is primarily responsible for forest dieback, whether it is only one of several contributing factors, or whether it plays little or no role is disputed at the present time. If it plays a major or contributing role, it is not clear which constituents are responsible or how they produce the damage. There are at least four viable but unproven hypotheses about how air pollution might adversely affect forests, of which acid rain is only one. The others postulate important roles for ozone, which has effects similar to those of SO_2 on plants, overfertilization by nitrogenous air pollutants (see Chapter 14), and stress, which represents the combined effects of all pollutants and perhaps natural disease.

SUMMARY POINTS

Even before the appearance of man on earth, millions of tons of sulfur per year have moved in a natural cycle from the atmosphere to oceans and land, then back again.

Volcanos and natural gas contribute H_2S and SO_2 to the atmosphere; these are converted to H_2SO_4 and sulfate salts that descend to earth with particulates or water.

Sulfate is taken up by plants; plants are consumed by animals; after death the wastes are acted upon by bacteria to make volatile forms of sulfur for return to the air.

Probably H_2S is one of these volatile forms, but at least five other compounds have been suggested as also making contributions to the atmospheric pool.

The burning of coal by humans now puts as much sulfur into the atmosphere in some areas as the natural cycle, and it is all in the form of SO_2.

Air pollution in the industrialized northeastern U.S. is of the reducing type, involving SO_2; the adverse lung effects of SO_2 are greatly potentiated by particulates.

Acid rain in that area is responsible for the destruction of natural food webs in lakes and ponds; it may also be responsible for large areas of forest dieback.

BIBLIOGRAPHY

Amdur MO, Corn M: The irritant potency of zinc ammonium sulfate of different particle sizes. Am Ind Hyg Assoc J 24:326, 1963.

Hofmann DJ: Increase in the stratospheric background sulfuric acid aerosol mass in the past 10 years. Science 248:996, 1990.

Johnson, NM, Reynolds RC, Likens GE: Atmospheric sulfur: its effect on chemical weathering of New England. Science 177:514, 1972.

Jorgensen BB: A thiosulfate shunt in the sulfur cycle of marine sediments. Science 249:152, 1990.

Kellogg WW, Cadle RD, Allen ER, et al.: The sulfur cycle. Science 175:586, 1972.

Lovelock JE, Maggs RJ, Rasmussen RA: Atmospheric dimethyl sulphide and the natural sulfur cycle. Nature 237:452, 1972.

Natusch DFS, Klonis HB, Axelrod HD, et al: Sensitive method for measurement of atmospheric hydrogen sulfide. Anal Chem 44:2067, 1972.

Nihlgard B: The ammonium hypothesis—an additional explanation to the forest dieback in Europe. Ambio 14:2, 1985.

Powell MA, Somero GN: Blood components prevent sulfide poisoning of respiration of the hydrothermal vent tube worm. Science 219:297, 1983.

Pratt, CJ: Sulfur. Sci Am 222:62, 1970.

Thom NG, Douglas RT: A study of hydrogen sulphide levels in the geothermal areas of Rotorua, New Zealand. Proc 4th Int Clean Air Cong Jpn, 565, 1977.

14

CHAPTER

THE NITROGEN CYCLE AND OXIDIZING TYPE AIR POLLUTION

Like sulfur (see Chapter 13), elemental, diatomic nitrogen (N_2) can be both oxidized and reduced. It, too, can exist in an unusually large number of stable valency states. With the exception of hydroxylamine, all of the chemicals listed in Table 14–1 play prominent roles in the natural nitrogen cycle.

NITROGEN AND LIFE

The earth's atmosphere contains 79% N_2 gas by volume. The French chemist, Lavoiser, who was the first to arrive at that figure, gave to N_2 the name, azote, meaning lifeless. His reason was that N_2 alone will not sustain plant or animal life. These need the 21% of the atmosphere consisting of oxygen. As scientists were eventually to realize, however, Lavoisier's view was too narrow because both plants and animals require vast amounts of nitrogen to sustain life. Inadequate supplies of nitrogen may be the most common limitation to biologic growth. When such a limitation occurs with respect to plant life, it importantly affects human hunger. Therefore, N_2 is in the anomalous position of being one of the most abundant of the elements and at the same time a scarce nutritional resource.

The reason for the above paradox is that the most abundant chemical form of nitrogen, namely N_2, is chemically and biologically inert. It cannot be used as such by plants or animals. In order to make nitrogen available to plants or animals, it must be changed chemically by a process known as fixation (Fig. 14–1). Fixation is accomplished by combining nitrogen with other elements, particularly oxygen or hydrogen. In the natural nitrogen cycle, fixation is carried out by bacteria (see below), but scientists have discovered how to fix nitrogen synthetically. As much as 25% of all nitrogen fixation is now carried out synthetically to make fertilizer by the so-called Haber process. In this process atmospheric N_2 is made to combine with hydrogen gas (H_2) to form

TABLE 14–1. THE STABLE OXIDATION STATES OF NITROGEN

Formula	Name	Nitrogen Valency
$NaNO_3$, HNO_3	Sodium nitrate, nitric acid	+6
$NaNO_2$, HNO_2	Sodium nitrite, nitrous acid	+3
N_2	Nitrogen	0
H_2NOH	Hydroxylamine	−1
NH_3	Ammonia	−3

ammonia gas (NH_3). Ammonia may be used as such for fertilization or it can be converted to other, more stable forms, such as nitrates or urea. Most nitrogen fixation (75%) is carried out by a few species of bacteria and algae. No higher living species has the natural capacity for fixation although some plants help in the process indirectly (see below).

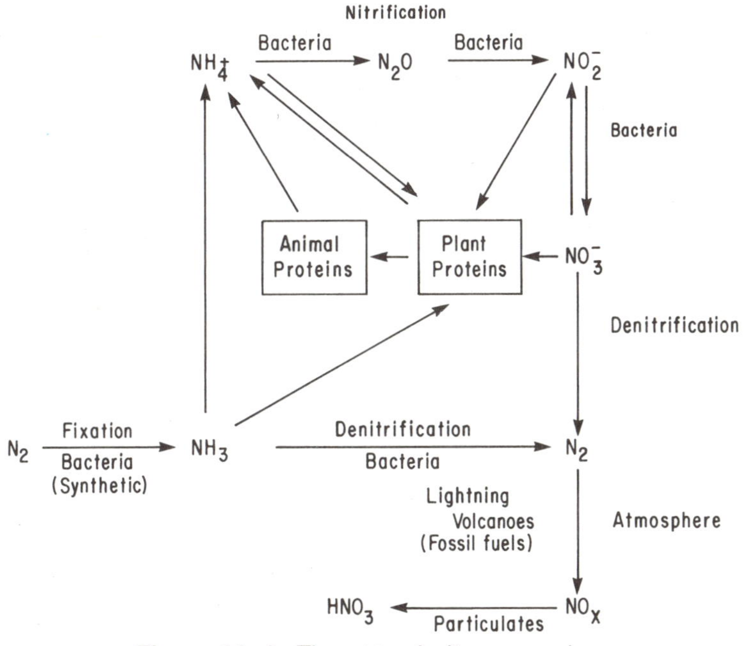

Figure 14–1. The natural nitrogen cycle.

THE NITROGEN CYCLE

As in the case of sulfur (see Chapter 13), one must evoke a volatile nitrogenous compartment in the cycle in order to return it to the atmosphere. In this case the volatile link is known with certainty to be the same as the starting material, N_2. The process for converting fixed nitrogen back into N_2 is called denitrification (Fig. 14–1). It is also carried out primarily by bacteria, but different ones from those involved in fixation.

Before the era of modern agriculture, one could be confident that the amount of N_2 removed from the atmosphere by natural fixation was balanced by its return to the atmosphere as a result of denitrification. Today, some uncertainty exists regarding whether or not denitrification is keeping pace with the sum of natural and synthetic fixation. Nor can anyone predict the long-term consequences of such an imbalance. One such consequence might be increased eutrophication of ponds and lakes. Excessive runoff of nitrogenous fertilizers may cause algeal blooms, as has been alleged for the phosphate detergents (see Chapter 15).

Minor processes for fixation include the burning of fossil fuels, such as gasoline or coal, and lightning discharges and volcanic eruptions, some of which generate sufficient heat or energy to fix atmospheric N_2. Blue-green algae in oceans, bacteria that live freely in water or soil and bacteria that have formed symbiotic relationships with plants are the microorganisms that carry out natural fixation. The best known of the latter are the bacteria of the genus *Rhizobium,* which live in nodules on the roots of legumes such as soybeans, peanuts, alfalfa, beans, peas, and clover. It has been known for centuries that legume crops revitalize soil that has been depleted of fixed nitrogen by such crops as corn. This is the reason for crop rotation in which fields are planted in alternate years with legumes and grains. Additional fertilizer must still be added to maximize grain yields, but the amount needed is less when the strategem of rotation has been used.

Soils can actually be inoculated with commercially grown *Rhizobium* bacteria to ensure that the best possible relationship evolves when the legumes are planted. The bacteria are taken up by the roots along with nutrients from the soil by specialized hair cells. The bacteria multiply within the root in a manner resembling an infection, and the roots form nodules containing bacterial colonies. Although the process resembles a disease, it is actually beneficial because it results in N_2 fixation for the nutritional needs of the plant. The different *Rhizobium* species show a high specificity for various legumes, often invading only one or two species of plants. Thus, the bacteria specific for peas may not invade alfalfa. A few nonlegume plants also form symbiotic relationships with N_2-fixing bacteria.

A certain parallelism exists between bacterial fixation and synthetic fixation. First, the stable triple bond between the two atoms of nitrogen must be broken, and then three atoms of hydrogen must be added to each to form two molecules of ammonia. These reactions have a high energy cost, and in the synthetic process the energy is supplied by burning fossil fuels. Thus, the price of fertilizer is directly related to the price of coal or petroleum.

In the soil NH_3 is converted quickly to ammonium ion (Fig. 13–1), which can be assimilated by the plant and used to make proteins. Alternatively, bacteria may act on ammonium ion to convert it to nitrous oxide, a volatile gas used in surgery as an anesthetic (so-called "laughing gas"). The nitrous oxide (N_2O) is in turn converted to nitrite and nitrate, which are forms that also can be assimilated by plants. As in the sulfur cycle, when the plants are consumed by animals, the plant protein is broken down and reincorporated into animal proteins. When the plants or animals die or give off waste products, fixed nitrogen is returned to the soil for recycling or denitrification. Similar reactions occur in ocean waters.

FIXED NITROGEN AND FERTILIZERS

Up to about 1900 the major source of fixed nitrogen both for agriculture and for the munitions industry were large deposits of sodium nitrate in Chile. As these deposits neared total depletion, the prospect of impending scarcity of nitrate fertilizer and its potentially devastating effect on world food supplies provided the incentive for the development of the Haber process. As fossil fuels are gradually depleted, yet another source for fixed nitrogen may have to be sought. Perhaps improvements can be made in the efficiency of the natural process by screening both plants and bacteria to find the best combinations. There are thousands of species of known legumes, and there may be an infinite number of mutant forms of *Rhizobium*. It is possible that genetic engineering might allow the transplantation of the nitrogen fixing apparatus, the nitrogenases, into other species of bacteria or even directly into plants that could be made self-fertilizing. It had long been thought that nitrogenases required the metal, molybdenum, for nitrogen fixation, but nonmolybdenum nitrogenases have recently been identified and may offer new alternatives for exploitation in agriculture.

In addition to pesticides (see Chapter 9), fertilizers constitute a second broad category of agricultural chemicals. They are available in many different physical forms and in many different chemical formulations. The nutrients needed in largest amounts (macronutrients) by plants are nitrogen, phosphorus, and potassium. Secondary elements include sulfur, calcium, and magnesium. There are an additional seven essential trace elements or micronutrients. Fertilizers are identified by an N-P-K code. For example, formula 8-32-16 contains 8% fixed nitrogen by weight, 32% phosphorus pentoxide equivalent by weight, but in some other chemical form such as superphosphate, and 16% potassium oxide equivalent, but in the chemical form of potassium chloride or sulfate. Adding the total yields 56 fertilizer units. The mixtures are never 100% because of other added materials. Formula 16-20-0-14S contains no potassium, but it has 14% elemental sulfur by weight.

The above types of fertilizers have a low order of acute toxicity, but certain specialized fertilizers can be dangerous. Anhydrous ammonia gas liquified under pressure is applied directly to soils or mixed with irrigation waters. Like SO_2, NH_3 is a lung irritant that in high concentrations can produce immediate airway obstruction and death. Highway accidents involving tank trucks and train tanker derailments have produced human poisonings and death. A new technology, particularly in midwestern states involves the processing of large volumes of livestock wastes as liquid manure. Manure is a good source of fixed nitrogen, but many gases are generated in the process. The most dangerous of these is hydrogen sulfide, and several deaths among exposed farmers have occurred (see Chapter 13).

In accordance with modern agricultural practice vast amounts of fixed nitrogen fertilizer are applied to soils in midwestern and western states and in the corresponding provinces of Canada. Irrespective of the form in which it is applied it is eventually converted in the soil to nitrate salts. Unlike DDT, nitrate salts are highly water-soluble and do not persist in soil. Nitrate runoff to ground waters and to deep aquifers is appreciable in farming communities. So much so, that in many parts of the midwest the nitrate concentrations in potable water supplies often exceed the Public Health Service recommendations. The problem intensifies during the growing season. Fortunately, the common nitrate salts are only moderately toxic to humans and animals. In man, low doses of sodium nitrate have no biologic effect. They are rapidly absorbed and excreted unchanged. In larger doses they have mild laxative and diuretic effects.

NITRITE IN THE ENVIRONMENT

Problems arise with nitrate in the environment because it is so readily converted to the much more toxic nitrite. The conversion occurs under a variety of circumstances, some of which are not well understood. If bacteria are not the most commonly involved mechanism for that conversion, they are certainly the most commonly blamed. Nonpathogenic bacteria capable of reducing nitrate to nitrite are found growing normally in the human gastrointestinal tract. These bacteria, however, only come in contact with ingested nitrate under special and unusual circumstances.

In nonruminant animals nitrite salts ingested orally are at least 10 times more acutely toxic than the corresponding nitrate salts. In all animals nitrite has two well known adverse effects. It directly relaxes blood vessels causing them to dilate, and, therefore, the blood volume is inadequate to fill the enlarged space. Blood flow to the brain is decreased along with its oxygen supply, and the victim faints. Fainting is actually autoprotective particularly if the feet end up higher than the head because that restores the blood flow to the brain, and the person may recover consciousness. The second toxic effect of nitrite is that it reacts chemically with the hemoglobin in the blood to convert it to methemoglobin. Methemoglobin is dark greenish brown instead of the bright red of normal hemoglobin when combined with oxygen. If it is generated in high enough concentrations in humans, it will impart a blue-gray color (cyanosis) to the skin, lips, and nail beds. A mini-epidemic of nitrite poisoning occurred in a New York Bowery soup kitchen when sodium nitrite was confused with ordinary table salt. The incident was described in an essay, entitled "Eleven Blue Men." Paradoxically, it is this effect of nitrite that makes it a useful antidote in cyanide and sulfide poisoning (see Chapter 3). Both of these toxic effects of nitrite are shared by hydroxylamine (Table 14–1), a chemical that does not participate importantly in the nitrogen cycle.

As already indicated, nitrates are inert in most animals, but ruminants (e.g., cattle, goats, and sheep) represent an important exception. These animals have a four-chambered stomach consisting of the rumen, the reticulum, the omasum, and the abomasum. The grass they eat is swallowed unchewed, and it passes into the rumen or the reticulum. Eventually, it is regurgitated and chewed as a cud. When the cud and saliva are swallowed, they pass through all four chambers where they are acted upon by gastric juices. In this special anatomic case any nitrate ingested along with the food is exposed to normal bacterial growth in the rumen. In man and other animals the nitrate is absorbed before reaching the habitat of the normal bacteria in the intestines. In ruminants, however, there are both opportunity and time for the bacteria to convert nitrate to nitrite and for the nitrite to be absorbed and exert its characteristic toxic effects. For this reason ruminants are almost as sensitive to nitrates as they are to nitrites.

In 1945, the first case in what was to become an epidemic of well water methemoglobinemia (infantile methemoglobinemia or the blue baby syndrome) was reported in Iowa. A survey conducted in 1949 and 1950 turned up 278 cases with 39 deaths. However, reporting was not mandatory for this disease, and the true number of cases may have been substantially larger. Most of the U.S. cases occurred in midwestern states, but similar cases were reported in Canada, England, Belgium, and Germany. In the U.S., reports of new cases fell off steadily through the early 1950s without any corrective action having been taken knowingly. Today the disease has all but disappeared with only rare, sporadic reports appearing in the literature.

Typically, the reaction occurred in infants under 1 year of age. Often the child was a newborn and brought home from the hospital in apparent good health. After a highly variable period of time ranging from days to weeks, cyanosis would be noted. Eventually, the entire infant might turn a slate gray color and drawn blood specimens would be brown or black. The child was usually taken back to the hospital, where it might be placed under observation, and often it made a gradual, spontaneous recovery. Occasionally, the above scenario would be repeated a second or third time. It was obvious that the infants were being exposed to a toxic chemical in the home environment, and when they were returned to the hospital, the exposure was interrupted. During their stay in the hospital, the red blood cells could repair the damage and convert methemoglobin back to normal hemoglobin.

The family always lived on a farm or in a rural area. No cases were ever traced to municipal water supplies. Older siblings, parents or others in the household were not affected. The water supply was usually a dug well, which meant that the source was not far below the land surface. When chemical analyses were performed, the water invariably contained high concentrations of nitrate, usually more than 40 ppm of nitrate nitrogen. Only 2% of all cases were associated with water containing less than 20 ppm and none with less than 10 ppm. Levels of more than 1000 ppm have been measured in some Kansas ground waters, and some municipal water supplies have more than 45 ppm, the upper limit recommended by the U.S. Public Health Service. In addition to fertilizer runoff, animal wastes from feed lots contribute to this nitrate nitrogen.

Some wells were found to be contaminated with bacteria capable of converting nitrate to nitrite. If, however, the well water had been the source of the nitrite, older members of the family drinking the same water would have been affected in at least some cases. The next link would be between the well and the infant. In that era it was fashionable to feed infants powdered formula which was mixed with the well water. In Czechoslovakia, infant formula contaminated with spores of nitrate-reducing bacteria was blamed for a serious outbreak of 314 cases. Although similar contamination was shown in some U.S. cases, it could not account for a large proportion of the total cases.

The remaining possibility was that something unusual happened in newborn infants that did not happen in older children. The methemoglobinemia was commonly associated with a gastrointestinal upset, vomiting and diarrhea. This is so common among newborns that it did not elicit much attention, but affected infants also tended to have a much lower gastric acidity (higher pH) than normal newborns or older children. In 1948, two physicians decided to conduct an experiment that today would probably not be permitted on ethical grounds. They were treating an infant who had been operated on for intestinal obstruction, and for legitimate reasons had a colostomy, an artificial opening from the abdomen to the bowel so that fecal material could be evacuated. They introduced sodium nitrate into the bowel through the colostomy, and within a few hours the child was methemoglobinemic.

Within the context of what was known about ruminants (described above) a plausible hypothesis was formulated that appeared to fit most of the known facts. The lower gastric acidity in affected infants permitted the nitrate reducing bacteria normally present in the lower bowel to survive higher in the gastrointestinal tract than they usually do. This circumstance provided the opportunity for the bacteria and the ingested nitrate to come into contact with each other. Nitrate was reduced to nitrite, and nitrite was absorbed to produce the illness. Gastric acidity normally rises during the first year of life, so if the infant survived a year, the illness was self-curative. This explanation is widely accepted today, but there are still puzzling aspects to the disease. One of these is the question of why there are so few cases in modern times when nitrate exposure not only continues but may be intensifying.

NITROGEN OXIDE GASES

Plants that are grown in nitrate rich soils tend to accumulate it above and beyond their needs for growth. There have been occasional reports of methemoglobinemia in infants consuming foods prepared with nitrate rich vegetables. Corn stalks accumulate high concentrations of nitrate, and the first reports of methemoglobinemia in cattle involved herds that had been fed cornstalks high in nitrate. It is common practice on dairy farms to chop together cornstalks, hay and straw to make a winter feed for cattle called fodder or silage. This "salad" is stored in silos, and for a short period of time after storage the silage "works." Powerful chemical reactions occur, and acids such as nitrous acid are generated along with toxic gases such as nitric oxide (NO) and nitrogen dioxide (NO_2); the nitrogen oxides are collectively referred to as NO_x. The silo reaches a high temperature and, occasionally, explodes. After several days the reaction subsides, and the danger period is past.

Exposure to high concentrations of the yellowish-brown fumes of NO_x during the danger period, however, as with SO_x or NH_3, can result in lung injury, pulmonary edema, and death. A rash of such cases in the 1950s led to the name, silo fillers' disease. Occasionally, the exposure has been so intense that NO_x was absorbed systemically to produce methemoglobinemia.

NITRATES

Another source of exposure to nitrates in our diet are the so-called processed or cured meat and fish. The origins of this process antedate recorded history. Probably ordinary salt was used as a preservative for meats at first, but when salt contaminated with nitrates was used accidentally, the superiority of nitrate was recognized. Pickling brines now contain salt peter or potassium nitrate, but the active agent is probably nitrite generated by bacterial action.

Curing has three desirable effects in such products as ham, bacon, franks, bologna, and pastrami. It is responsible for the characteristic reddish-pink color, which is caused by a reaction of nitrite with myoglobin, a close relative of hemoglobin that is found in muscle. Without curing, these meat products would be an unappetizing greenish color. Curing is also responsible for the characteristic flavor, the chemistry of which is still unknown. Finally, it has an antibacterial action that the meat-packing industry insists is uniquely effective against the growth of *Clostridium botulina,* the bacteria that produces the most poisonous substance known to man (see Chapter 3). The industry position is that curing is necessary to maintain the generally excellent record of safety of American meat and fish food products. They express the fear that the current low incidence of botulism from contaminated foods would increase greatly in the absence of curing. Although the FDA apparently does not find this argument convincing, they have not been successful to date in their efforts to ban the process. The present standards set a limit of 200 ppm nitrate/nitrite nitrogen in processed meat products, and surveys of U.S. products show most to be within those limits.

In 1954 in England, three workmen were exposed to a new and potentially useful solvent that had not yet undergone toxicity testing. Two of the men developed severe liver damage, and one of them eventually died. The solvent was dimethylnitrosamine (DMNA), and it is the simplest member of a family of hundreds of chemicals known collectively as nitrosamines (NA). Two years later DMNA was shown to be highly carcinogenic, and since then the NA have been the subject of intensive investigation.

More than 300 different NA derivatives have been tested for carcinogenic effects and about 95% have tested positive. Many of these are mutagenic and teratogenic (see Chapter 20). The carcinogenic NA have many different target organs, and examples are known for more than a dozen different organs and tissues. The organ specificity is influenced by all the factors known to influence toxicity in general (see Chapter 3). They have been tested for carcinogenic effects in more species than any other single class of chemicals: no less than 39 ranging from snakes and fish to monkeys. A resistant species has yet to be found. Many NA are extremely potent, and single doses of 1 mg/kg of certain NA induce tumors. All carcinogenic NA require biotransformation via hepatic microsomal enzymes for activation (see Chapter 7). Because they induce tumors in tissues with low levels of cytochrome P450 isozymes, it is presumed they are activated in the liver and transported by the blood to those tissues. At least some of the active forms of NA are also inactivated by P450 isozymes so whether or not a given NA is carcinogenic may depend on the relative rates of activation and inactivation. The active forms are genotoxic initiators that react directly with DNA (see Chapter 20).

In 1978 in Omaha, Nebraska, a jealous former employee of the Eppley Institute for Cancer Research bizarrely murdered two people and severely poisoned two others. The murderer had stolen a bottle of DMNA from a laboratory of the institute and poured it into a pitcher of lemonade in his former girlfriend's refrigerator. All of the victims had signs of severe liver injury.

The island of Guam in the South Pacific is part of the Mariana group. The seeds from a palm-like tree in the cycad family are a staple in the native diet, even though they are widely recognized to be toxic. The seeds are prepared as food in a special way that is believed to remove the toxic ingredient. Even so, the natives of Guam have a high incidence of several diseases in comparison with the rest of the world. Among these are cirrhosis and cancer of the liver. One of the toxic substances found in cycad nuts is called cycasin, and it has a chemical structure similar to that of DMNA. Considerable speculation exists regarding whether or not incomplete removal of cycasin during preparation of the nuts might be responsible for the high incidence of liver disease.

Few readers of this volume will ever eat cycad nuts, but the first suggestion that NA might be something more than a laboratory curiosity was evidence to suggest that NA might be formed in tobacco smoke. Subsequently, some was found in high concentrations in batches of nitrate-preserved fish meal. The alarming thing about NA is the ease with which they are formed. If nitrite is mixed with any of a number of so-called secondary amines (H-N-R_2), which are found in abundance in the environment and in our bodies, the reactions proceed spontaneously at neutral pH and room temperature. Thus, the suspicion is raised that cured meats and fish, which also contain high concentrations of amines, might be sources of NA. Indeed, they have been detected in such products, although the concentrations in general have been low. Fairly high levels, however, can be generated under certain conditions such as the frying of bacon. There are formidable analytical problems in measuring some NA, particularly those that are volatile, so it has been difficult to assess the degree of risk. Air pollution with NO_x might conceivably be another source of NA exposure.

Even if preformed NA are not being ingested in significant amounts, the possibility exists that they might be formed in our own gastrointestinal tracts as a result of the dietary content of nitrate and nitrite. Nitrate or nitrite mixed with amines and segments of isolated intestines from rats led to the formation of detectable amounts of NA. Mice dosed with both nitrite and amines showed an increased incidence of liver damage and a higher mortality than control groups given neither or given either alone. Several commonly used drugs can be converted to NA under certain conditions. The risks of this type of endogenous exposure are no clearer than are those for exogenous exposure.

OXIDIZING-TYPE AIR POLLUTION

Five chemicals or classes of chemicals are said to account for nearly all urban air pollution: 52% as CO, 18% as SO_x, 12% as hydrocarbons, 10% as particulates, and 6% as NO_x. As already noted the picture changes from location to location: smelters emit SO_x, steel mills generate particulates, and automobiles exhaust CO, hydrocarbons and NO_x. Los Angeles, with its enormous number of automobiles, has a different kind of air pollution problem than New York. It is called the oxidizing-type of pollution or photochemical smog. In addition to an inversion, photochemical smog is worsened by the ultraviolet radiation of intense sunlight.

The chemistry of oxidizing-type air pollution is more complicated than that of the reducing type. The major constituents are ozone, NO_x, aldehydes, peroxyacetyl nitrates, and hydrocarbons. All of these are toxic, except hydrocarbons, although they interact with atmospheric pollutants to generate the more exotic oxidants. Thus, as with reducing-type air pollution, photochemical smog is more hazardous than any one of its ingredients. Exposure to photochemical smog results in severe eye irritation, but evidence for other adverse effects, including excess deaths (see Chapter 13), has been slow to accumulate. One large-scale study in California compared an area of low pollution with an area of high pollution. The study was based on questionnaires and measurements of lung function. The populations were subdivided into cigarette smoking categories of never, light, and heavy. The expected differences between smokers and nonsmokers were found in both study localities, but the residents in the high pollution area were consistently in poorer health than those in the low pollution area whether they smoked or not. Smoking and air pollution seemed to have additive effects, and when the results were corrected for the smoking factor, the effect of air pollution still was statistically significant.

The bulk of the oxidant activity of photochemical smog is caused by ozone, and only about 10% is caused by NO_x. Ozone is a normal constituent of the atmosphere that increases in concentration from about 0.1 ppm at sea level to 10 ppm in the upper atmosphere. In the upper atmosphere it screens out certain kinds of ultraviolet radiation that are responsible for causing skin cancer. There is much concern today that atmospheric pollution with chlorofluorocarbons is resulting in a depletion of the protective ozone layer. At high altitudes it is generated by the action of sunlight on oxygen. It has occasionally reached toxic concentrations in commercial airliners flying above 30,000 feet. We have already seen that NO_x is also a normal constituent of the earth's atmosphere as a result of volcanic activity and lightning. Most of this is in the form of NO_2 at 0.1 to 0.15 ppm, with the concentration of NO about one third of that. The major sources of atmospheric NO_x, however, are anthropomorphic from the burning of coal and the operation of internal combustion engines.

As the temperature of burning gasoline increases, the hydrocarbon emissions decrease but the NO_x emissions increase. The reason for this is that at higher temperatures not only is more NO_x generated from nitrogenous constituents of gasoline, but additional NO_x is contributed from the fixation of atmospheric nitrogen. Field studies in California indicate that the introduction of the catalytic converter into automobile exhaust systems has reduced the emissions of CO, lead, and particulates, but doubled the emissions of NO_x. The atmospheric concentrations of NO_x have shown a generally increasing trend. The current national air quality standard of 0.05 ppm ($\mu g/m^3$) is regularly exceeded in Los Angeles, and if the present trends continue, the same will be true for Chicago and Philadelphia.

The toxic effects of ozone have much in common with NO_x. Ozone is the more reactive chemically of the two, whereas NO_x is the more water-soluble. For these reasons ozone tends to produce more superficial injury to the respiratory tract and the epithelial cells on its exterior surface, whereas NO_x tends to produce deeper more slowly developing lesions.

Although the atmospheric concentrations of NO_x are not so high as those of SO_x, the former does make a contribution to acid rain, and the trends with respect to relative concentrations seem to be reversing. As indicated in Chapter 13, one of the still viable theories about how acid rain might affect forest dieback is overfertilization with fixed nitrogen.

SUMMARY POINTS

Nitrogen also moves in a natural cycle between the earth and the atmosphere; it is one of the most abundant elements, but also can be a scarce nutritional resource.

Nitrogen cannot be used by plants or animals; it must be fixed by combining it with H_2 or O_2; this is done by *Rhizobium* bacteria living in nodules on the roots of legumes.

Man intrudes on the natural cycle by synthetically fixing large amounts of nitrogen to make fertilizers; the energy for this comes from the burning of fossil fuels.

SUMMARY POINTS (continued)

Fertilizer runoff leads to high nitrate concentrations in potable water supplies in agriculturally intensive areas where it may be converted to the more toxic nitrite.

High nitrate levels in water and plants are a cause of infantile methemoglobinemia, silo fillers' disease, poisonings of ruminants, and a potential source of nitrosamines.

The oxidizing-type of air pollution as in Los Angeles is associated with NO_x emissions from auto exhaust and intense sunlight; NO_x also contributes to acid rain.

BIBLIOGRAPHY

Binkerd EF, Kolari CE: The history and use of nitrate and nitrite in the curing of meat. Fd Cosmet Toxicol 13:655, 1975.

Brill WJ: Biological nitrogen fixation. Sci Am 236:68, 1977.

Delwiche CC: The nitrogen cycle. Sci Am 233:136, 1970.

Freussman, R: Carcinogenic N-nitroso compounds and their environmental significance. Naturwissenschaften 71:25, 1984.

Kohl, DH, Shearer GB, Commoner, B: Fertilizer nitrogen: contribution to nitrate in surface waters in a cornbelt watershed. Science 174:1331, 1971.

Maugh TH: Sulfuric acid from cars: a problem that never materialized. Sci 198:280, 1977.

Roueche, B: Annals of medicine, the prognosis for this patient is horrible. The New Yorker, Jan. 25, p. 57, 1982.

Smith, RJ: Ever so cautiously, the FDA moves toward a ban on nitrites. Science 201:887, 1978.

15

CHAPTER

THE DETERGENT CONTROVERSY

Robert E. Gosselin

In 1972, a prominent American commentator wrote: "One of the most persistent and bewildering national controversies of recent years has swirled around a matter that once attracted little public notice except in advertisements and nursery rhymes, namely, the way we wash our clothes." The great detergent controversy involved two health-related issues: the safety of detergent products for the consumer in the home and the safety of these products for the environment. It was the latter, the ecologic concerns, that captured the attention of the American public in the late 1960s and throughout much of the following decade. The affair is one of the first times that the general public on a nationwide scale became aroused about an issue primarily of ecology. Although the debate is no longer acrimonious, unresolved issues persist.

A HISTORICAL SURVEY OF SOAP AND OTHER SURFACTANTS

Before the 1930s the American home contained only a few kinds of cleaning agents. Washing soda and borax were available, as were various acid and bases used for cleaning metal and porcelain surfaces; and, of course, dry-cleaning and paint solvents. The only general purpose cleaning agents, however, were soap and water. The first appearance on the American market of a general purpose detergent other than soap occurred in 1933. Synthetic and semisynthetic detergents, however, did not cut into the soap market appreciably until the end of World War II, notably with the introduction of Tide in 1946. In spite of the bitter debate about detergents, their sales and presumably their consumption have continued to rise ever since. Today soap represents only a small part of the market for laundry products.

What is soap and what are the synthetic detergents that have so successfully replaced it? Figure 15–1 depicts the traditional way to make soap by the alkaline hydrolysis of neutral fat of animal

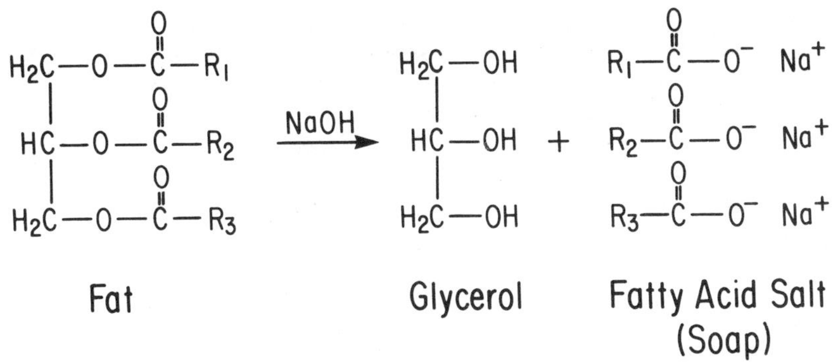

Figure 15–1. The manufacture of soap. Soap consists of a long linear chain of carbons with attached hydrogen atoms (R_1, R_2, and R_3) and a terminal acidic group (COO^-), which is charged and forms salts with sodium. (Na^+).

or vegetable origin. The chemist calls this reaction "saponification." The products are glycerin and the sodium salts of various "fatty" acids. These organic salts are then allowed to solidify to form bar soap. In modern industry, acid rather than alkali is used to hydrolyze fat, especially animal fat (tallow). The fatty acids released are isolated by steam distillation and then neutralized with sodium or potassium hydroxide (or a mixture of the two hydroxides). Thus, commercial soap consists chemically of sodium or potassium salts of fatty acids. Fatty acids are composed of straight (i.e., unbranched) carbon chains. There may or may not be unsaturated bonds in the chain, but in every fat of natural origin the number of carbon atoms is even. Frequently, the chain length lies between 12 and 20 carbons. Chemists use the symbol R to refer to these long straight "alkyl" chains.

The two ends of such long linear molecules have entirely different properties. The carboxyl group (Fig. 15–1), called the polar end partly because of the charge that it bears when it ionizes in water, has a high affinity for water. In contrast, the alkyl portion of the molecule is nonpolar; it finds no affinity for water but seeks fat and fatty solvents. As a result, molecules like soap tend to collect and orient at interfaces. Their presence at air-water surfaces reduces surface tension. Surface tension is the property that makes water drops on a smooth surface round up. In the presence of molecules that reduce surface tension, the drops tend to spread to form thin films. Interfacial tension between solid particles and water is also lowered so that the coalescence of suspended particles is prevented. At interfaces between immiscible liquids, the reduction of interfacial tension promotes emulsification. The alkyl end of soap molecules cluster around dirt particles, particularly greasy dirt particles, which detachs them from fabrics and causes them to be dispersed in water. Thus, soaps serve as surface-active agents or so-called surfactants. Because they acquire a negative electric charge when they ionize in water, they are classified as "anionic surfactants."

Soap is only one example of a surface-active molecule. Other surfactants are also useful cleansing agents. The word "detergent" is sometimes used to mean any surface-active molecule with cleansing activity other than soap itself. The first of the semisynthetic detergents to have a significant impact on the American market was introduced in 1933. It was made by reducing fatty acids obtained from vegetable oils (especially coconut) to the corresponding fatty alcohols; the alcohol group was then sulfated. The sodium salts are known as sodium alkyl sulfates. Like soap, these molecules have a polar and a nonpolar end. As long as the alkyl chain is only about 12 carbon atoms long, the alkyl sulfates have an important advantage over soap, namely a resistance to precipitation in hard water.

Other classes of surfactants have also been synthesized and found to have special uses. Cationic surfactants, so named because they acquire a positive electric charge when they ionize in water, are excellent protein precipitants. Cationics with single long alkyl groups are used as germicides and "sanitizers" on chemically clean, inert surfaces, such as counter tops and surgical instruments, but they are not found in laundry or general purpose household detergents. Cationics with two long alkyl groups (referred to as "ditallow" materials) are waxy and very insoluble in water; small quantities are sometimes included in laundry products to serve as a fabric softener.

The principal surfactants in modern detergent formulations, however, are anionic and nonionic. In granular laundry products the anionic is chiefly reponsible for the cleansing action. In liquid laundry products the surfactants are chiefly, if not exclusively, nonionic. Many of the nonionics are related to the common antifreeze, ethylene glycol. They are made by polymerizing ethylene oxide to form chains of which the repeating unit is ($-CH_2-O-CH_2-O-$). The reaction is conducted in the presence of fatty alcohols, which also react with ethylene oxide. As a result the alkyl group of the fatty alcohol becomes one end of the polymer. The product is called an "alcohol ethoxylate." Even though these molecules have no ionizable groups, they do have polar and nonpolar ends and, therefore, are surface-active.

During World War II the civilian population collected and donated table scraps and other sources of fat as a patriotic duty. This highly organized and effective campaign served to increase the supply of glycerin needed for the manufacture of TNT. In processing this fat for its glycerin content, a mixture of fatty acids was an inevitable and useful byproduct. A considerable surplus of fatty acids must have been on hand at the end of the war. Paradoxically, this is the time that the soap and detergent industry chose to introduce a variety of new synthetic and semisynthetic surfactants as substitutes for soap. Like soap itself, most of them were of the anionic type and all possessed an alkyl chain to form the nonpolar head. In some cases this alkyl chain was obtained from natural fatty acids available from tallow and vegetable oils. As noted before, however, sulfation of fatty alcohols yields a versatile product only if the alkyl groups are relatively short, such as those found in coconut oil. Unlike mixed animal fat, coconut oil was in short supply at the end of World War II. Therefore, manufacturers sought other sources of nonpolar residues for making surfactants, and increasingly petroleum hydrocarbons were used, including both straight-chain and cyclic (benzene) moieties.

To replace the sulfation of fatty alcohols, the industry began to sulfonate alkyl benzenes. The result was a proliferation of alkyl benzene sulfonates (ABS). One particular example of an ABS depicted in Figure 15–2 has been named sodium tetrapropylene benzene sulfonate. From the late 1940s to about 1963, it was probably the most widely used anionic surfactant in American detergent products. Like soap itself and the earlier alkyl sulfates, one end of this molecule is obviously anionic and the other is nonpolar. The nonpolar residue here was produced by polymerizing propylene derived from petroleum. ABS possesses a carbon-to-sulfur bond (Fig. 15–2). This bond is more resistant to hydrolysis (i.e., rupture) than is the carbon-to-oxygen-to-sulfur linkage in the alkyl sulfates (Fig. 15–2). Consequently, the sulfonates are stable and effective in very hot water, and so they can be used in what the industry calls "heavy duty detergents." There are, however, some disadvantages. Unlike the alkyl sulfates, sodium alkyl benzene sulfonates share with soap the undesirable property of precipitating as calcium salts in hard water. The industry's solution to this problem was one of far-reaching significance, as we will see later.

Still another characteristic of the benzene sulfonates was not recognized immediately, namely their high chemical stability in the environment. Thus, bacteria and other microorganisms in the soil and water do not readily decompose these compounds, as they do fatty acids and fatty alcohol sulfates. In regions of the U.S.A. with ground water levels close to the surface, the benzene sulfonates began to accumulate in subsoil and surface water. Especially in areas lacking sewage disposal plants or central water reservoirs, low concentrations of surfactants appeared in the drinking water. As a result, when one turned on the faucet, he obtained a glassful of foam. The singer-satirist Tom Lehrer de-

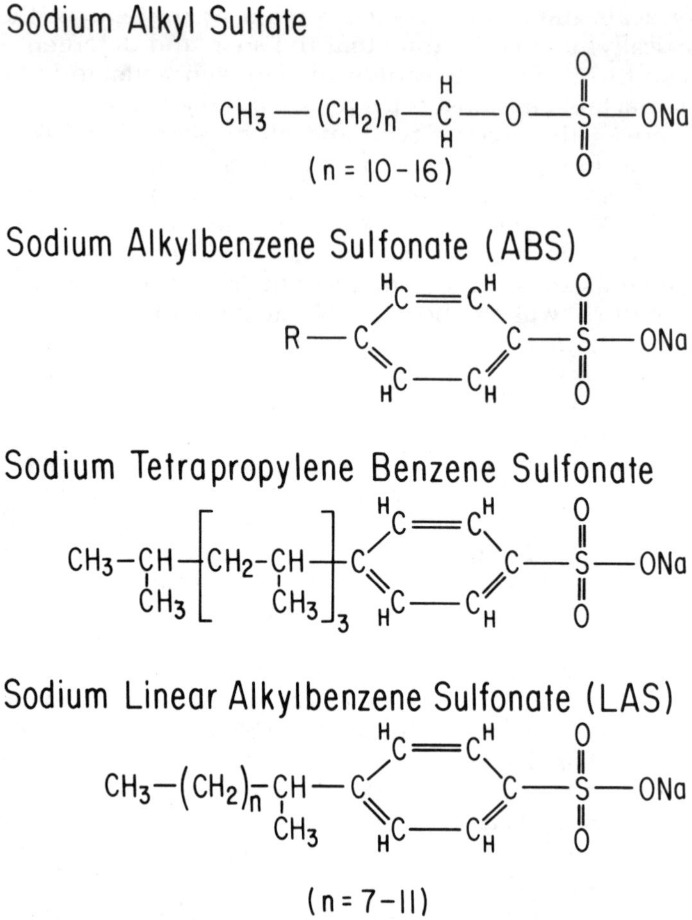

Figure 15–2. Anionic surfactants. Note the chemical difference between sulfates with a -C-O-S-bond and the sulfonates in which the sulfur is bonded directly to the carbon (-C-S-). The R group may be a straight or branched chain of carbons, and the cyclic structure is a benzene ring.

scribed it as "hot and cold free-flowing crud." Almost everyone found this experience outrageous.

In the mid 1960s Suffolk county in eastern Long Island responded by passing a law banning all synthetic detergents. This county had and still has a particularly difficult problem. In spite of expensive homes, many communities there have no centralized water supply or sewerage facilities; in these districts homes are serviced by individual artesian wells and septic tanks. With such a shallow water table the cycle between sewerage and well water is sometimes unpleasantly short in terms of time and distance. Based on the sudsing, many wells were tested and found to have dangerously large numbers of intestinal bacteria. Unlike the bacteria, a trace of surfactant in the drinking water is not a health threat, but it is unesthetic and, therefore, unacceptable.

The detergent industry responded quickly and effectively to the gathering storm clouds of public indignation. Within one year no more branched chain anionic surfactants were marketed. All were replaced with biodegradable substitutes, so-called because they are rather quickly broken down by soil and water microorganisms. For the most part these substitutes belong to a class of compounds known as "linear alkyl sulfonates," labeled LAS in Figure 15–2. As you can see, they are still derivatives of benzene sulfonate, but the alkyl chain is now unbranched. Frequently, the chain consists of 12 or 13 carbon atoms, but it is derived from petroleum, not from fatty acids. Since 1964–1965, when the branched chain compounds were completely replaced, foaming and sudsing in wells, ponds, and lakes seem to be under good control. No significant environmental problems have been traced to the linear alkyl sulfonates, which are still widely used.

Anionic and most nonionic surfactants are also essentially hazard-free from the viewpoint of consumers in the home. Fed to test animals in large quantities, they cause bloating and diarrhea but little else. In chronic, high-level feeding studies some interference with nutrition may occur because they stimulate gastric mucus secretion and inactivate at least one digestive enzyme. No systemic toxic effects, however, are recognized, perhaps because these compounds are not appreciably decomposed within or absorbed from the gastrointestinal tract. When applied to the skin, especially during prolonged exposures, they are described as mildly "defatting," and so produce some roughness and reddening. Sensitization and allergic reactions, however, are rare. In terms of their ability to produce irritation of the skin and mucous membranes, anionic surfactants, presumably because of their mild alkalinity, are generally more active than nonionics, but in the eyes, some of the nonionics are more irritating. In any case significant injuries are extremely rare.

DETERGENT BUILDERS: EFFECTIVENESS AND SAFETY OF POLYPHOSPHATES

Surfactants are not the only ingredients in modern laundry detergent products. Indeed, they constitute considerably less than

50% of the usual dry granular formulations. Table 15–1 describes some typical compositions, some old and some current. Many kinds of additives are present. Some, like sodium silicate in small amounts, are designed to inhibit corrosion in modern washing machines. Others serve as antiredeposition agents to emulsify dirt and prevent its resettling on the fabric, especially when the wash water is spun out of the clothes. Some components function as anticaking or antidusting agents. Brighteners are usually included, as well as perfumes, and sometimes bacterial proteolytic enzymes.

The bulk of these formulations, however, is made up of ingredients known as "builders." The builder is so named because it extends the usefulness of the surfactant by preventing its inactivation or precipitation by elements responsible for water "hardness." In most natural waters the chief offenders are ionic calcium and magnesium. In the absence of effective builders all of

TABLE 15–1. APPROXIMATE PERCENTAGE COM DETERGENT

Ingredients	Function	Light Duty Liquid Detergent
Surfactants	Detergency: removal, dispersion and emulsification of dirt	
Anionic (usually sodium salt)		
Linear alkyl benzene sulfonate		
Alkyl sulfate		5–9
Alkyl ethoxysulfate		15–33
Non-ionic		
Alkyl ethoxylate		
Alkylamine oxide		1–4
Builders	To chelate or precipitate elements of water hardness	
Sodium tripolyphosphate		
Tetrasodium pyrophosphate		
Sodium carbonate		
Sodium sulfate	Also serves as a formulation aid	
Additives*		
Sodium silicate	Corrosion inhibitor with some builder action	
Dialkyl dimethyl ammonium chloride	Fabric softener	
Liquids	Solvent in liquid preparations	
Water		50–74
Ethyl alcohol		5–9

*May also include perfumes, fluorescent dyes (as optical brighteners), carboxymethyl each at less than 1%, and mild oxidants (for bleaching) at less than 10%.

(After Booman, Cascieri TM, Demetrulias J, et al: In vitro methods for estimating eye the sulfonates, like soaps, tend to form greyish, greasy curds. In the case of soap we are familiar with this precipitate as a ring around the bathtub. If a similar precipitate were allowed in washing machines, it would adhere to the fabrics and be difficult to wash out. Furthermore, over any long period of time it would clog up any modern automatic washer. Precipitated surfactant is wasted because it cannot function as a cleansing agent. Even if one starts with soft water, an average load of dirty clothes contains sufficient calcium and other elements to be equivalent to water of moderate hardness. Thus, detergent builders appear to be indispensible.

Since 1946, when Tide was first marketed, the principal builders in detergent formulations have been the sodium salts of various inorganic polymers of phosphoric acid. The backbone of these molecules is a sequence of alternating oxygen (O) and phosphorus (P) atoms. Linear polymers of various lengths can be manu-

POSITION OF SOME REPRESENTATIVE MODERN PRODUCTS*

Liquid Laundry Detergent	Liquid Laundry Detergent	Granular Laundry Detergent (Nonphosphate)	Granular Laundry Detergent (Low Phosphate)	Granular Laundry Detergent (Phosphate)
	25–49	10–15	5–9	1–4
			5–9	5–9
			5–9	5–9
10–24	10–24			
			10–24	25–49
		25–35		10–24
		40–50	25–49	25–49
		5–10	10–24	1–4
1–4				
50–74	50–74	3–5	5–9	5–9
10–24	5–9			

cellulose (to prevent redeposition of dirt), proteolytic enzymes (as stain removers),

irritancy of cleaning products. Phase I: preliminary assessment. J Toxicol 7:173, 1988.)

factured, and even two-ring polymers have been synthesized. Mixtures of some of the longer polymers are used in industrial and home water softener formulations, but in laundry detergents only the trimer and occasionally the dimer are used. They are called sodium tripolyphosphate and sodium pyrophosphate, respectively. Tripolyphosphate, in particular, forms highly stable and soluble complexes with calcium and heavy metals in solution. This process, which is called chelation, protects anionic surfactants from precipitation.

What is known about the toxicity of these phosphate polymers? All of them form aqueous solutions that are mildly alkaline. When ingested, however, solutions of sodium tripolyphosphate are sufficiently irritating in the stomach to induce prompt vomiting, usually of rather short duration. Clearly such vomiting protects against the possibility of systemic poisoning, but in case vomiting does not occur, one would like to know something about the actions of these materials when present in the blood stream.

When moderately high doses are injected into test animals, two biochemical effects have been detected, either of which might compromise life. The first effect is rather obvious. When infused intravenously into rats, for example, polymeric phosphate chelates calcium in the blood stream and reduces the level of free ionized calcium to the extent that the cardiac action can be impaired. The heart may stop beating completely if the dose is large and the rate of administration is fast. If the animals survives, as it usually does when the material is administered by any route other than the intravenous one, a second toxic effect becomes manifest. It arises from the presence in various tissues of poorly characterized enzymes that hydrolyze the linear open chain polymers to orthophosphoric acid. Thus, even though the original solutions are mildly alkaline, their hydrolysis products are strongly acidic. The result is a metabolic acidosis that seriously compromises the body's buffer reserves.

All of these considerations, however, are somewhat theoretical because the phenomena of hypocalcemia and acidosis do not occur when material is administered by mouth or by stomach tube. Even if vomiting is not induced, only small amounts of these materials are absorbed into the blood stream, at least in rats and rabbits, and presumably in humans. Admittedly, some orthophosphate (i.e., nonpolymeric phosphate) is absorbed; probably it is produced by intestinal bacteria that degrade the polymers. To obtain a dose equivalent to the mean lethal dose in rats, a 2-year-old child would have to swallow and retain without vomiting about half a cupful of a typical granular phosphate detergent. Such a feat has never been accomplished. According to reports by poison centers, however, thousands of toddlers try every year; they sample their mother's detergent, mouth it and in some cases swallow a few granules, but no instances of metabolic acidosis or hypocalcemia have been established, presumably because the dose was too small or because the exposure was terminated by prompt vomiting. Indeed no deaths or substantial personal injuries have been traced to the phosphate detergent formulations that have dominated the consumer market for the past 45 years. Few household products enjoy such an excellent record of safety.

If safe and demonstrably effective as cleaning agents, why were these detergents such a subject of controversy? The explanation can be found in the word "eutrophication." In many American homes it became a battle cry in the late 1960s, but only a few years before, the word was known only to a few ecologists and limnologists.

EUTROPHICATION

Eutrophication refers to the process wherein natural bodies of water become enriched with nutrients to the extent that they support excessive growth of algae. Algae are free-floating microscopic photosynthetic plants, often referred to as phytoplankton. In moderate amounts they are an important part of the food chain for aquatic animals, but in some highly enriched lakes and ponds they proliferate to the extent of producing a thick blanket of green slime. As they die and sink to the bottom, bacterial decomposition occurs, and this consumes some of the oxygen dissolved in the water. In shallow ponds the oxygen tension may be reduced to the point where fish can no longer survive. In any case the water acquires a bad taste and a bad smell and is no longer suitable for many recreational purposes.

Algal blooms of this kind occur in the summertime only in those bodies of water that are particularly rich in nutrients. The process of eutrophication appears to be widespread and increasing. If one flies over the lake districts of Wisconsin and Minnesota in August, one has the impression that all lakes and ponds are covered with green scum. The problem is not so severe in New England, but algal blooms do occur. In the western basin of Lake Erie, which is the most eutrophic part of the Great Lakes, reports describe a green or yellowish green or bluish green velvety blanket over two feet thick that covers about 800 square miles of lake surface almost every summer. Obviously the problem deserves serious attention.

What are the nutrients that cause these algal overgrowths? More than 20 substances have been shown to be required by algae, including such elements as iron, molybdenum, magnesium, sodium, potassium, silicon, manganese, biotin, and niacin, but most attention has settled on the requirements for carbon, nitrogen, and phosphorus. With respect to carbon, carbon dioxide, or carbonate anion is sufficient for these photosynthetic algae. Diverse forms of organic nitrogen can be utilized; some of the blue-green algae can fix atmospheric nitrogen (see Chapter 14). Phosphorus is required in the form of orthophosphate.

Beginning about 1967, various scientists inside and outside the federal establishment began to point the finger of suspicion at phosphate as the rate-limiting nutrient for algal growth in the Great Lakes. Whether this claim is true or not, it does serve to remind us that a great deal of phosphate is washed into the aquatic environment every year. Household laundry detergents are one important source. Of course, only polymeric phosphate is present in detergents, particularly tripolyphosphate, but as soon as this material enters the natural environment, microbial enzymes break it down rapidly into orthophosphate. Over 2.5 million metric tons of detergent are consumed each year in the United States, representing more than a million metric tons of phosphate. But detergents are not the only significant source of phosphate that enters our waterways; animal and human excreta are rich in phosphates, as is the rain runoff from fertilized agricultural fields and the leaching of natural phosphate deposits. In municipal waste water, about half of the phosphate is estimated to come from detergents, but storm sewers and the runoff of rain into waterways also contribute phosphate, so that the percentage from detergents is less than 50% in most places. In any case the amount is large.

Essentially all experts appear to agree on three assertions about eutrophication.

1. It is not a new phenomenon. Green Bay, Wisconsin, was named by American Indians long before synthetic detergents or chemical fertilizers were employed. Many scientific studies of lake eutrophication antedate the availability of phosphate detergents.
2. Sunlight is essential for algal blooms. No matter what the nutrient content of the water, algal overgrowth is more apt to occur in clear shallow ponds than in deep turbid lakes.
3. Only ponds, lakes, estuaries, and slow moving rivers retain enough nutrients to sponsor algal blooms. No eutrophication is likely to occur in rapidly flowing rivers or in most ocean coastal waters.

Based on these assertions at least one environmental scientist has concluded optimistically that 85% of the U.S. population live in areas where phosphate in waste water is unlikely to promote eutrophication. For example, it has been estimated that 25% of the population live in rural areas and are served by septic tanks. If these households have adequate leach beds and typical ground water levels, phosphate in their sewerage never reaches waterways because of the high capacity of most soils to adsorb and retain prodigious amounts of this chemical. Some 55 to 60% of the population live in and around cities where waste water is discharged into rapidly flowing rivers or oceans. According to these estimates, only 15% of the population live where there is a eutrophication problem or where one might arise. Such a propitious assessment, however, is not widely endorsed.

How good is the evidence that phosphate is the rate-limiting nutrient in areas where algal blooms do occur? This question has proved to be surprisingly difficult to answer. Ecologists, limnologists and others do not even agree on the test procedures that might yield definitive answers. One experimental approach has been to take lake water into the laboratory and to study the growth of naturally occurring algae under artificial sunlight, with and without the addition of minute quantities of chemically pure nutrients. On the basis of such tests, nitrogen has been implicated as the rate-limiting nutrient in Lake Tahoe and in the shallow marine waters off the southern shore of Long Island. Indeed, high concentrations of sulfate in sea water are known to inhibit the uptake of molybdenum by algae; without this element (as molybdate), phytoplankton have an impaired ability to fix nitrogen or assimilate nitrate (see Chapter 14). Silicates are reputed to be a limiting nutrient in the eutrophication of some of the New York State finger lakes, and at least in a few soft water lakes and ponds, a deficit of available carbon is thought to be critical.

It is undoubtedly true, however, that phosphate is the rate-limiting nutrient in many natural bodies of water. For example, when phosphate was added continuously for 26 days to a tundra river in northern Alaska, effecting an enrichment of 10 parts of phosphorus per billion parts of water, a prompt growth of attached algae was found more than 7 miles downstream. Similarly, phosphate is believed to be the critical nutrient in the open, nutritionally deficient waters of Lake Superior, Lake Huron, and Lake Michigan. Almost certainly this is not true in the two eastern Great Lakes, Erie and Ontario, which are already highly eutrophic and in which the phosphate concentration is already so high that its availability does not appear to limit algal growth. Something else must be limiting, because the amounts of available phosphate could allow much heavier algal blooms. Although neither Lake Erie nor Lake Ontario can be properly described as "dead lakes," as is sometimes done in the lay press, both are badly polluted with substances considerably more toxic than algae. Eutrophication is only a small part of the problem.

Can phosphate removal convert a nutrient-rich lake into an oligotrophic one that will not support algal blooms? The answer is "probably no." The magnitude of the problem is suggested by the fact that as little as 10 micrograms of phosphate phosphorus per liter is sufficient to support good algal growth. Many parts of Lakes Erie and Ontario already possess considerably higher concentrations. Moreover, mud at the bottom of Lake Erie has been sampled by cutting a core 30 feet deep. A uniform phosphate content was found at all levels, suggesting a fairly steady laying down of phosphates, a process which began long before the introduction of synthetic detergents. Of course, it may be unnecessary to remove phosphates from lake bottom silt in order to reduce concentrations in the water, because there is no evidence that this sedimented phosphate redissolves to an appreciable extent. In any case the Great Lakes deserve the serious attention of all thoughtful persons, if only because they constitute the largest body of fresh surface water in the entire world (see Chapter 18).

INDUSTRY, GOVERNMENT, AND POLITICS

Reliable information about eutrophication was not widely available to the general public in the 1960s and 1970s. Until recently, lay concern about water quality focused on eutrophication and the role of phosphate. To solve what was perceived as an inherently simple problem, many sincere and dedicated persons organized campaigns to ban phosphate detergents. For example, in 1970–1971 a complete ban on the sale and use of phosphate detergents was instituted in Erie County, New York (which includes Buffalo), in Dade County, Florida (which includes Miami), and in Cook County, Illinois (which includes Chicago). Since then the ban has been extended to the entire states of New York and Indiana and to the District of Columbia. As recently as 1985, Maryland joined the ranks of states outlawing conventional phosphate formulations. Many other states, counties, and cities passed laws reducing the allowable levels of phosphates in detergents, pending a complete ban. In terms of population, about one-third of the United States is now covered by ordinances banning or restricting the sale and use of phosphate detergents.

The detergent industry did not respond in a single-minded way to this new challenge, as it had earlier to the public concern about nonbiodegradable surfactants. Indeed, the three giants of the industry (Procter & Gamble, Lever Brothers, and Colgate-Palmolive) hesitated. Many smaller companies, however, promptly marketed laundry products that were phospate-free and allegedly "nonpolluting." Because these phosphate-free formulations rapidly captured about 10% of the market, the giants of the industry lost some of their competitive advantage.

The solution adopted by the smaller companies and by several new companies that quickly appeared was to substitute in place of polyphosphates one or more builders that precipitate calcium. For this purpose they returned to some of the oldest known washing materials, notably washing soda, which is sodium carbonate, to sodium silicate, and in some cases to sodium metasilicate. Essentially, all advertisements for these products at that time stated that they were "environmentally safe" "nonpolluting," or "unpolluting," but no evidence was presented to support such assertions. Because algae require both silicates and carbonates, it is reasonable to infer that these phosphate-free detergents might promote algal growth in some natural bodies of water. The principal objection to the new detergent formulations, however, was the issue of safety in the home. Carbonates, silicates and metasilicates are effective only in solutions of rather high alkalinity. Some of the new products were so alkaline, as we will see later, that informed physicians regarded them as household hazards. Apparently this same concern was in large measure responsible for the reluctance of the big three detergent manufacturers to join the antiphosphate crusade.

In any case, the largest of the big three, the Procter & Gamble Company, was convinced by the summer of 1970 that it had an acceptable solution in terms of a new, phosphate-free builder that is readily biodegradable and does not require high alkalinity. Sodium nitrilotriacetate is widely known by the initials NTA. Over a period of several years, NTA had been subjected to considerable laboratory testing, both for efficacy and safety, as well as some environmental testing. By mid 1970 it was already being used in small test markets in partial replacement of tripolyphosphate in some P&G laundry products.

On December 18, 1970, the United States Surgeon General abruptly requested the industry to suspend voluntarily the use of NTA pending further testing. His request was based on a preliminary report from the National Institute of Environmental Health Sciences, which indicated that NTA may have teratogenic effects in laboratory rats when combined with various heavy metals, presumably by enhancing the transmission of these metals across the placental barrier into the fetus. Because this study could not be confirmed in subsequent years by either governmental or industrial scientists, the prohibition of NTA in detergents was lifted several years later.

The ban on NTA in December, 1970, left the consumer with only two choices: to continue to purchase the conventional phosphate detergents where available or to embrace the new, highly alkaline, phosphate-free materials. Because government spokesmen had played an important role in molding public opinion against phosphates, it was widely assumed that they preferred the newer materials. Nine months later, on September 15, 1971, a memorable press conference was held in Washington, D.C. It was attended by Surgeon General Dr. Jesse Steinfeld, Head of EPA William Ruckelshaus, Head of the President's Council on Environmental Quality Dr. Russel Train, and Head of FDA Dr. Charles Edwards. The message was that NTA would not be cleared quickly for renewed use, but that the public should realize that some of the products being marketed as phosphate substitutes were sufficiently alkaline to constitute a hazard to the user. Dr. Steinfeld even suggested that for the time being housewives might prefer phosphate detergents because they are "safe for human health." Many dedicated ecologists were outraged and regarded this press conference as a sellout to the big three. However one interprets the testimony, it is clear that the federal establishment did not exert any strong or consistent leadership in this area. The confusion of the public was not difficult to understand.

THE EFFICACY AND SAFETY OF NONPHOSPHATE BUILDERS

By 1971–1972 most manufacturers had replaced part or all of the polyphosphate in their laundry detergents with various mixtures of nonphosphate builders. These low-phosphate and phosphate-free formulations generated many consumer complaints because laundered clothing was not as white or bright as formerly. Some homemakers are said to have made regular trips across state and county borders in order to purchase conventional high-phosphate products. Because the latter are still banned in many parts of the U.S.A., it is probable that this informal bootlegging continues even today, but perhaps the public has become more tolerant of what was once called "tattletale gray."

The nonphosphate builders that replaced polyphosphates were various mixtures of sodium carbonate, sodium silicate, and sodium metasilicate; in some cases sodium bicarbonate and sesquicarbonate were also included. As noted above, they act by precipitating calcium, not by chelating it, but they are effective only in solutions of moderately high alkalinity. This alkalinity, that so concerned Dr. Steinfeld and others, is readily demonstrated by titrating solutions of pure builder with standardized solutions of any mineral acid.

Figure 15–3 depicts titration curves of sodium carbonate, sodium metasilicate, and sodium tripolyphosphate. In each case the concentration is 1%, which is perhaps 20 times higher than would be found in a washing machine under usual conditions, but is considerably lower than might be accidentally ingested or splashed in the eye. Each curve illustrates two essentially independent indices of alkalinity. The first index is the solution pH before the addition of acid. Figure 15–3 shows that sodium carbonate and sodium metasilicate have much higher initial pHs than tripolyphosphate. Concentrated solutions are, of course, still more alkaline, but the pH of even saturated solutions of sodium tripolyphosphate never exceeds 10.5, whereas the pH of strong sodium metasilicate solutions lies above 13.

It is generally agreed that pH is a measure both of the intensity of a solution's alkalinity and of its potential for inducing tissue

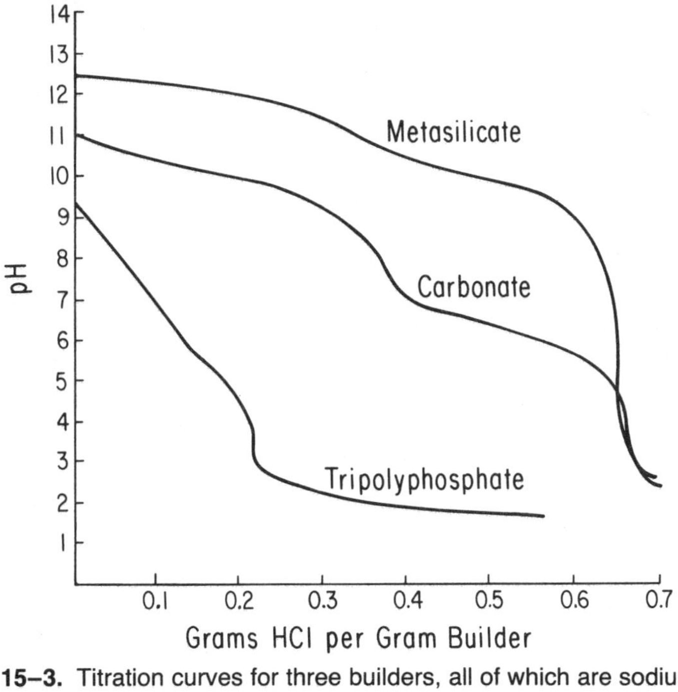

Figure 15–3. Titration curves for three builders, all of which are sodium salts. (See text for details.)

injury, provided that the solution can sustain its pH when tissue constituents begin to neutralize it. This pH sustaining power, or so-called buffer capacity, is the second index of alkalinity and of the potential for producing alkali burns in living tissue. A common measure of buffer capacity is the amount of acid required to lower the solution pH to neutrality (pH 7); this quantity is called the "titratable alkalinity." Figure 15–3 demonstrates that sodium carbonate and especially sodium metasilicate have much higher titratable alkalinity than tripolyphosphate. The titratable alkalinity, however, is of concern only if it represents buffer capacity in a pH range that is high enough to be dangerous. For chemical burns of the mouth and esophagus, any pH above about 10.5 is potentially hazardous. According to Figure 15–3 the titratable alkalinity in this range is about 4, 0.5, and 0 (grams of hydrochloric acid per gram of builder) for metasilicate, carbonate, and tripolyphosphate, respectively. Concentrated solutions have much higher values, except for tripolyphosphate where the solution pH does not rise above 10.5.

Not surprisingly, detergent products containing these builders tend to share the titration characteristics of the builders themselves. Figure 15–4 shows titrations performed in the author's laboratory in 1971 by several concerned Dartmouth College students. Each curve represents a 1% solution of a different granular

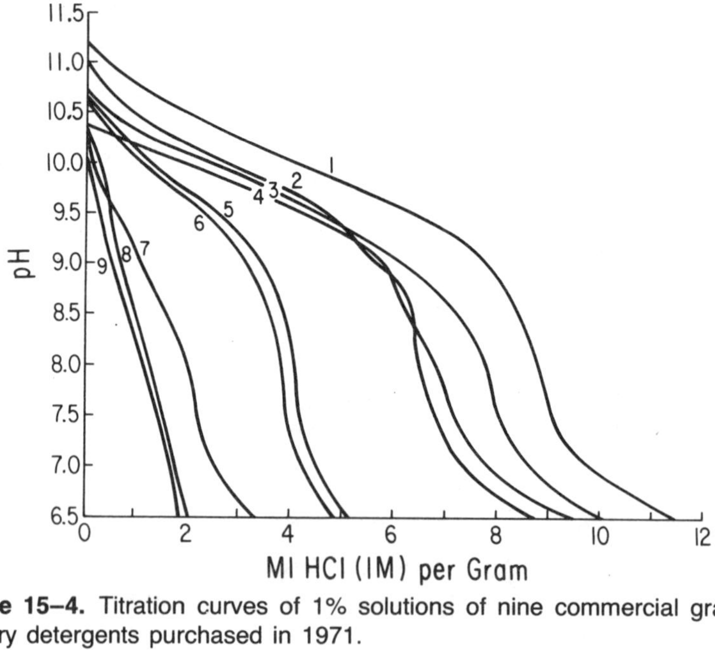

Figure 15–4. Titration curves of 1% solutions of nine commercial granular laundry detergents purchased in 1971.

laundry detergent bought at the local supermarket. (The brands are not named here because we believe that many of them are formulated differently today and so have different titration curves.) Brands 1 through 6 were advertised as "phosphate-free," brand 7 was composed of soap granules without builders, and brands 8 and 9 were polyphoshate formulations typical of that era. The nonphosphate products may also have been typical of that time, but we know that even more intensely alkaline brands were marketed.

The public had little or no warning that the "new" laundry detergents, which were abruptly introduced into the consumer market, were so much more alkaline than the phosphate detergents they replaced. The good safety record of the phosphate products during a quarter century of home use may have inspired a false sense of security. As soon as alerted to the new changes, the federal government threatened to require precautionary health warnings on detergent package labels. Toddlers, however, are not dissuaded by labels, and whether because of ignorance or innocence or carelessness, accidents do occur in the home.

The expected injuries were of the same type that had long been associated with exposures to lye, which was and still is found in many drain cleaners, paint removers, oven cleaners, and some automatic dishwasher detergents. Most of these injuries consist of chemical burns of exposed surfaces, particularly the cornea of the eye and the mucous membrane linings of the mouth and esophagus. The ingestion of lye often produces prompt vomiting, but this vomiting is hardly autoprotective because the highly sensitive esophageal lining is then re-exposed to the corrosive action of the ingestate. Even if injured mucous membranes finally heal, scarring of the esophageal wall may lead to such severe strictures that swallowing becomes impossible. Clinical experience indicates that any product with the capacity to sustain its pH above 10.5 (i.e., with appreciable buffering above pH 10.5) should be regarded as hazardous. With a series of such products, the intensity of tissue damage rises sharply as product pH exceeds 11.5. Because concentrated solutions of some of the original nonphosphate laundry detergents had a pH as high as 12, public health officials predicted a rash of severe alkali burns.

SEQUEL

What has happened to the "great detergent controversy" since the early 1970s? Laundry detergents are still produced and consumed in quantities of 3.5 million metric tons annually. Zero-phosphate products are still marketed, but they have not gained ascendency over phosphate detergents, as many observers expected. Algal blooms still occur; whether they are less intense than formerly has not been established, but some improvement has been described in isolated areas.

As expected, the zero-phosphate products did cause some severe alkali burns and at least one death of a toddler. The predicted epidemic of injuries, however, does not seem to have occurred. The explanation appears to be that the most dangerous brands were quickly withdrawn from the market or reformulated to be less caustic. Specifically, sodium metasilicate was deleted in most instances and replaced with such builders as sodium carbonate, silicate, and aluminosilicate. The lowered alkalinity reduced the washing efficacy of these products, but the changes in composition forestalled the imposition of federal regulations that would have required stringent warning labels.

The big three detergent manufacturers eventually marketed their own versions of zero-phosphate products. One and the same name brand is now available in several different formulations, each marketed in a different region of the country to match local ordinances. For example, at least three versions of Tide are manufactured, two with phosphate and one without. The phosphate-free, low-alkaline builder NTA is still produced and extensively used in detergent products in Europe and Canada but not the U.S., perhaps in part because New York instituted a statewide ban on it.

A clear, industrywide trend has been to reduce the level of polyphosphate in both laundry and dishwashing detergents, even in products with no other added builders. This technologic change has been accomplished by the development of surfactants that are less sensitive to calcium and other minerals in hard water. For the most part, however, the newer anionic surfactants, like the original alkyl sulfates, are suitable only for light duty detergents. Liquid detergent formulations, which include some heavy duty laundry products, have been developed in the past decade. For the most part they do not contain and do not require builders because their detergency is based on calcium-insensitive nonionic surfactants, rather than calcium-sensitive anionics (see Table 15–1). Anionics, however, are still the most widely used and effective surfactants, and so builders cannot be entirely eliminated.

That this country or any other modern industrial society will support a complete ban on polyphosphates is probably unrealistic, as is also the conviction that a complete ban will serve to eliminate eutrophication. We now recognize that phosphate is only part of the eutrophication problem and that detergents are only one source of environmental phosphate. The detergent controversy cannot be legislated out of existence. The issues are real and warrant imaginative and constructive solutions.

One effective way to reduce any environmental damage from excessive amounts of phosphorus, whether derived from detergents or fertilizer, is to process municipal waste water through a tertiary sewage treatment operation. For example, treating the effluent from conventional sewage plants with alum or lime would serve to precipitate not only phosphate but also many other algal nutrients; it would also serve to remove some other kinds of pollutants, such as many toxic heavy metals. This process has long been recognized as technologically feasible, and many regard it as economically sound, at least in geographic areas at major risk. Of course expanding sewage treatment plants usually requires tax money. Many legislators prefer to express their environmental sensitivities by banning phosphate detergents rather than by raising taxes. Most communities do not have an adequate tertiary sewage treatment process in operation.

Although the great detergent controversy of the 1960s and 1970s may appear in retrospect to have been a quaint episode in the course of human affairs, it cannot be dismissed as a trivial one because it served, as no other issue before it, to mobilize public interest and opinion in environmental problems generally. Certainly the detergent debate is no longer a vigorous one, not because the public has become tolerant of eutrophication of its lakes and ponds but because other aspects of environmental pollution are now viewed as more critical, notably contamination of surface and ground water by pesticides and other synthetic chemicals of high toxicity and long persistency (see Chapter 18).

The detergent controversy was a great teacher. One lesson that should have been learned, but probably was not, is that the home is an important part of our environment and that sound ecologic principles should apply indoors as well as outdoors (see Chapter 16).

SUMMARY POINTS

Soaps are made from animal fats and vegetable oils. When dissolved in water, soap molecules ionize to form anions that accumulate at surfaces, as do some synthetic polymers made from petroleum. All of these materials are called "anionic surfactants," and they are present in most laundry products.

SUMMARY POINTS (continued)

Before World War II, soap was essentially the only surfactant used in laundering clothes. From the late 1940s to 1963, the most widely used anionic surfactants were alkyl benzene sulfonates (ABS) made from petroleum. ABS molecules proved to be stable and, therefore accumulated in waste water and in soils. In areas like Long Island, NY, where the ground water table is close to the surface, detergent foam began to appear in tap water. Although this foam was not a health hazard, it was unesthetic and the public was outraged. When linear alkyl sulfonates (LAS) were substituted for ABS, this problem quickly disappeared because soil bacteria readily decompose LAS.

In hard water all soaps and most other anionic surfactants precipitate as calcium and magnesium salts. The resulting greasy curds have no cleansing power, and if allowed to form, they would clog modern washing machines. To prevent this, all granular laundry products contain "builders" that serve to protect the surfactants by binding calcium, magnesium, and other elements of water hardness. Since the mid 1940s, the chief builders in laundry products have been sodium salts of polymeric phosphate, notably sodium tripolyphosphate.

Like the surfactants, polyphosphates have proved to be safe and effective in the home. Discarded in waste water, they are broken down by soil microorganisms into inorganic phosphate. The latter is one of many nutrients required for the growth of simple aquatic plants known as algae.

The presence of algal overblooms in lakes and ponds is called "eutrophication." Because of water odor, taste, and fish kill, it is a significant and apparently growing environmental problem.

Although detergents supply only part of the phosphate in ground water and phosphate is not always the rate-limiting nutrient in eutrophication, various states, counties, and cities passed laws in 1970 to 1972 banning or restricting the use of phosphate builders in detergent products.

In response, the detergent industry introduced phosphate-free, "nonpolluting" products in which silicates, metasilicates, and carbonates were substituted for polyphosphates as builders. Some of the early products, however, were so alkaline and corrosive that they were a hazard in the home, and they were eventually modified or withdrawn.

Today consumers can buy detergents that have conventional amounts of phosphate, low phosphate, and even no phosphate. Among the latter are liquid products that require no builders because they contain nonionic rather than anionic surfactants.

Several nationally distributed brands differ in composition in different parts of the country to meet local ordinances.

16

CHAPTER

INDOOR AIR POLLUTION

If the cleanliness of the air in our urban environments leaves something to be desired, one solution might be to remain indoors in a temperature and humidity controlled artificial atmosphere from which noxious pollutants from the outside air have been removed by filtration, electrostatic precipitation, and other means. This is one manifestation of the "space-ship mentality" which will soon be put to a test as a selfsustaining biosphere in the Arizona desert. It is estimated that urban residents already spend better than 90% of their time indoors. That figure might be even higher for the very young, the elderly and the chronically ill, particularly those with cardiovascular or respiratory diseases. Unfortunately, as seen in Chapter 13, these groups are particularly susceptible to airborne chemical or microbiologic insults.

Because of the amount of time spent indoors, more than one third of U.S. energy consumption occurs in buildings from the most modest of family homes to gigantic offices and stores. Of the approximately 100 million buildings in the U.S., about 82 million are residential, either as single family dwellings or apartment complexes. The increased cost of energy consumption to control indoor environments has led to the increased use of insulation and the repair of structural defects that might result in heat loss or gain, an increased use of caulking, and more tightly sealed storm doors and windows. These result in buildings that are much less well ventilated than in the past. Outside air infiltrates into homes and buildings through openings and cracks around doors and windows and areas cut out for pipes and wiring. Although this sounds like a relatively small amount, complete air turnover can occur every one to two hours in the average American home.

Especially in private homes patterns of fuel consumption have changed, and households have diversified to avoid over-reliance on a single source of energy. Sales of wood stoves increased about ten-fold throughout the 1980s, and the use of kerosene space heaters and coal stoves also increased dramatically. Thus, the sources of indoor air pollutants in the form of pyrolysis products are increasing at the same time that homes are becoming less well ventilated. Although there is considerable interest and experimentation with renewable solar and wind sources of energy, these cleaner technologies have not been exploited to their full potential for various reasons. Gas and electricity are cleaner than coal or fuel oil, but they are not available to all homes at competitive prices. With respect to outside air, however, it makes a great deal of difference whether the electricity is generated by the burning of coal or by hydroelectric means.

Surprisingly, hospitals account for 15% of all energy consumed in commercial buildings, and hospitals and their associated laboratories contain many chemicals. Any reduction in the rates of air exchange in such buildings may represent a trade-off between savings in energy expenditure and increased exposure to airborne chemicals or infectious agents. Huge and complicated ventilation systems are commonly installed to comply with building codes for adequate ventilation. A hospital serves many different functions and the ventilation codes appropriately take the functions into account. For example, some generally accepted figures for "once-through" complete air turnovers per hour is 20 for surgical suites, 10 to 15 for laboratories, and 5 for offices. Elaborate systems of heat exchangers allow part of the energy expended in heating or cooling exterior air for admission to the interior to be reclaimed before it is vented outside again. Air conditioning systems are not without their hazards in that they are sometimes the source of infectious bacteria growing in pools of water in the duct work (see below).

Because of the escalation in housing and energy costs, many U.S. families are turning to smaller homes such as prefabricated units and mobile homes. In the U.S., mobile homes constitute about 20% of the new housing market. These may be even less well ventilated than more conventional homes, and they are apt to contain more of the newer synthetic materials, which emit volatile pollutants. The more important hazardous materials found in indoor air include tobacco smoke, particulates, carbon monoxide, nitrogen dioxide, sulfur dioxide, carbon dioxide, formaldehyde, benzene, methylene chloride, chlordane as used for termite control, chloroform, asbestos fibers, airborne allergens, and infectious microorganisms. The most widely publicized incident of the latter was the outbreak of deadly Legionnaire's disease, which was spread through the air conditioning system in a Philadelphia hotel in 1976. Of the 153 documented cases, 29 resulted in death. Because the focus of this book is on toxic chemicals no more will be said about infectious agents and allergens, but these problems are often aggravated in the winter when people spend a larger proportion of their time indoors. The phenomenon of winter allergy is a counterpart to the seasonal, outdoor, allergic challenges of pollens.

The problem of indoor air pollution was first recognized in public office buildings when such a large number of workers complained of similar symptoms that the situation could not be ignored by supervisors. Ironically, the EPA's Washington headquarters was picketed in 1988 by 70 aggrieved workers who wanted improvements in the indoor air quality. Worker complaints in that and similar incidents took the form of stuffy heads, watery or burning eyes, nosebleeds, headaches, general malaise, fatigue, dizziness, rashes, metallic taste in the mouth, changes in contact lens color, and difficulty in breathing. After repeated episodes in numerous buildings were publicized, few authorities doubted that the complaints had a legitimate basis, but it was not possible to prove that the symptoms were causally related to exposure to any particular chemical or group of chemicals. So, the constellation of signs and symptoms came to be known as "the sick building syndrome."

The systematic study of the sick building syndrome is in its infancy, but the few available preliminary reports have been interesting to say the least. One 5-year study was sponsored by the EPA and conducted by the Research Triangle Institute in North Carolina. Dr. Edo D. Pellizzari and his associates monitored 350 people in Elizabeth and Bayonne, NJ, by measuring and comparing their exposures to 20 different common chemicals at home, outdoors, and in the workplace. In their preliminary report the levels of 11 of the 20 were as much as 10 times higher in the home environment than nighttime levels outside. These included chemicals derived from petroleum and found in gasoline, paint and halogenated hydrocarbon solvents, such as tetrachlorethylene and trichlorethylene. Peak hourly NO_2 concentrations of between 200 and 700 $\mu g/m^3$ have been measured routinely in kitchens during conventional cooking on gas stoves, whereas the National Ambient Air Quality Standard is 100 $\mu g/m^3$. Proof of adverse health effects from such a source, however, is still lacking.

Surprisingly, the levels of indoor pollutants in a pristine community in North Dakota were on a par with those in the New Jersey homes. Although the investigators were unprepared for these findings, they noted that the indoor concentrations were still far below federal limits for exposure in the workplace, e.g., the TLV (see Chapter 4). Potential adverse health effects were not included in the study so the significance of the findings is unknown at present. However, it is obvious that if exposure in the home is greater than that outdoors, the regulatory agencies need to take the total exposure into account, including time-weighted averages, in drawing up guidelines for air quality standards. This realization opens a Pandora's box of policy issues. Whereas a precedent may exist for drawing up regulations for air quality standards in public buildings, no precedent and little enthusiasm exists for trying to regulate air quality in the home.

Estimates of the number of sick buildings, i.e., those that have areas in which 20% of the employees have complaints that are eased when they leave the premises, place the incidence at one-fifth to one-third of all existing buildings. The sources for the various chemicals above may not yet have been completely identified, but they probably include rugs, draperies, wall panels, copier machine fluids, carbonless paper, paint, and cleaning solvents. It is rare that a single chemical or source can be identified as the root cause. Most often the exposure is to chemical mixtures, and the alleged effects of the chemicals may be modulated by noise levels, lighting, design of work spaces and a host of other factors. The sick building concept has been extended to a group of people who have embraced identification as "building diagnosticians." These groups draw from the talents of experts in interior design, health, biology, and behavior to try to improve the indoor environment.

Concern about indoor air pollution became widespread in the 1970s when OPEC began threatening to ration or to increase the cost of exported oil. In the rush to conserve fuel, home insulating activity suddenly increased, particularly in the northeastern U.S. Polymeric foams of phenol-formaldehyde were in common use at the time. Numerous complaints of headache, nausea, eye and throat irritation, difficulty in breathing, insomnia, nosebleeds, and rashes began to be reported to health agencies. It was widely held that these symptoms resulted from the leaching of unpolymerized formaldehyde from the foam material in building walls and ceilings. In 1982 phenol-formaldehyde foams were banned in schools and residential buildings, but formaldehyde did not go away. Formaldehyde is found also in a number of building materials such as plywood, particle board, carpets, draperies, permanent press cloth, adhesives, and paint preservatives in the form of resins such as urea-formaldehyde. The half-life for outgassing from such materials is more than 4 years. Newer materials obviously give off more gas than older ones, and increasing humidity and temperature increase the rates of emission. In 1985 the Federal government set limits on formaldehyde emissions from plywood and particle board. It has been seriously suggested that new or remodeled buildings should be subjected to a "bake out" period in which both the heat and ventilation are turned up to drive volatile chemicals outdoors.

Formaldehyde is mutagenic in a variety of bacteria and a nasopharyngeal carcinogen in mice and rats. Human volunteers complained of eye irritation at 0.1 to 0.4 ppm, but complaints were received from residents of homes where the levels were only 0.02 ppm. It has long been suspected that an unusually wide range of sensitivities exists to these effects of formaldehyde among the general population, but the duration of exposure may also be a determinant of sensitivity. Long-term, low-level exposures in the home, may be worse than short volunteer exposures in chambers. The possibility that formaldehyde exposure may itself induce greater sensitivity has not been ruled out.

Although formaldehyde may have been the agent to attract attention to indoor air problems, enormous public reaction accompanied the news from The EPA that the air in perhaps as many as 10% of U.S. homes appeared to contain dangerously high levels of radon. Radon is a colorless, odorless, tasteless, radioactive gas that is generated naturally as a breakdown product of radium and other natural radioactive elements. Radium is found in soils, rocks, and certain building materials. Thus, its spontaneous decay in the soil immediately under or around buildings can result in the accumulation of indoor radon, or in the contamination of drinking water from wells in the vicinity. The highest levels are found in basements and crawl spaces where ventilation is usually poor, and the levels in winter are commonly 60% higher than those in summer also because of ventilation. Although certain geographic areas are recognized as being at higher risk for radon, considerable differences were found among individual homes in a neighborhood, sometimes as close as the house next door. Presumably these differences were caused by differences in the underlying deposits of radium and in the design and construction of individual homes, but it means that each home must be tested individually. Fortunately, if high levels are found, they can often be reduced to acceptable levels by such relatively inexpensive measures as installing ducts to vent the radon outdoors or sealing cracks in basement walls and floors.

No overt warning of exposure occurs which necessitates the use of special but relatively inexpensive radon-testing kits. Radioactivity is known to cause human cancer, and since radon is a gas, its inhalation results in lung cancers. As with SO_x, however, the radon is often adsorbed to the surface of particulates that may be deposited through the respiratory tract depending on their size. The EPA felt at one time that 20,000 cases of lung cancer annually in the U.S. were caused by radon exposure in the home. This estimate was based on data accumulated over a number of years (late 1940s to early 1950s) in studies on uranium miners of the Colorado Plateau. The smoking habits of these workers were known and documented, and it was shown that the risks for lung cancer in miners who smoked was increased further by intensive exposure to radon.

Unfortunately, the original study and the extrapolations that the EPA made from it contained many flaws. The risks were assessed by extrapolating from high exposure levels in the miners to low exposures in homes. Such extrapolation is common, but the assumption that it is valid has never been proved. The conditions in the mines at that time were primitive, cramped, and often unventilated. The study attempted to measure what must have been a fairly small effect of radon that was superimposed on a very large effect of smoking. This approach is always difficult, and there may have been other confounding variables as well. Revised estimates indicate that at the low levels found in the average home radon does not constitute a health threat. Individual home owners, however, are advised to invest in radon testing since prospective buyers may insist on it as a condition to purchase. In this instance the public was not well served by the EPA. The EPA asbestos program in public buildings may also have been ill advised in some cases.

Whereas formaldehyde may have been the first widely publicized indoor air pollutant and radon could have been the most dangerous had the original calculations proved to have been relevant, undoubtedly the most ubiquitous is sidestream tobacco smoke. The largest cause of lung cancer in the home is undoubtedly passive (sidestream) cigarette smoke. Fully one-third of the "adult" population of the U.S. regularly smoke cigarettes. The percentage of children living in homes with one or more smokers varies with socioeconomic status and other factors, but it may be anywhere from 50 to 75%. Like some of the commercial products previously described (see Chapter 11) tobacco smoke is such a complex mixture that it defies precise chemical description. More than 2000 distinct chemicals have already been identified, and doubtless many more remain to be found. Although the health effects of secondary smoke have not been studied with anything approaching the depth and intensity of the effects on primary smokers, former Surgeon General C. Everette Koop declared sidestream tobacco smoke as a significant health hazard, and the National Academy of Sciences has concluded that involuntary exposure ought to be minimized or avoided altogether whenever possible. This policy is most clearly evident today in airliners and restaurants.

A number of other substances are recognized as primarily originating indoors and are found inside homes in much larger concentrations than outside. Ammonia results from human metabolism as well as cleaning products. Pentachlorophenol (penta or PCP) is widely used as a wood preservative, and like formaldehyde it can leach from wooden materials and products for long periods after treatment. Additional exposure can occur if such wood is burned in stoves or fireplaces within the home. Penta is absorbed through the intact skin (see Chapter 9), but it is much more efficiently absorbed, perhaps as much as 100%, when the vapors are inhaled. It can cause bone marrow damage, and is said to be teratogenic, mutagenic, and carcinogenic as well (see Chapter 20). As indicated in Chapter 11, technical grade penta may be contaminated with dioxins and dibenzofurans. It has been established by body burden monitoring that measurements of penta and its derivatives in urine show a close correlation with environmental exposure.

The levels of NO_x, CO_2, CO and particulates, which are often much higher indoors than out, may have their origin in fuel burning either inside or outside the home. The indoor sources also include furnaces, chimneys, gas stoves, and auto exhaust from attached garages. Various volatile organic chemicals, some of which are low molecular weight halogenated hydrocarbons, may arise inside from adhesives, cooking, solvents and cosmetics, but outside sources are recognized as well. p-Dichlorobenzene originates from room deodorizers. Perchlorethylene residues can be readily identified as emanating from freshly dry-cleaned clothes, and chloroform can arise from chlorinated tap water.

Not all parties agree that intervention on behalf of indoor air pollution is justified, and even those who hold that position are not clear on what type of intervention is appropriate. As already noted above it is not a given that indoor air, particularly in private homes, is within the public domain. Public agency responses could take many forms such as leaving the issue to personal choice, which is the present state of affairs with respect to saccharin. Although cigarette smoking is also said to be a personal choice, the incredible social pressures that have been brought to bear on smokers place it in a special category of its own. The alleged problem could be ignored; educational campaigns could be mounted to effect change by persuasion; efforts could be expanded under existing legislation; fiscal incentives could be put in place; or new rules and regulations could be formulated. All of these, however, would be directed primarily at public buildings. Indoor air quality standards would be totally impractical and almost impossible to enforce in private homes where the major exposure may occur. On the other hand, it is no longer clear that the extremely costly measures already put in place and proposed for the future to control outside air pollution are in fact improving human health by reducing exposure.

SUMMARY POINTS

Because of increased energy costs, private homes are more tightly sealed, less well ventilated, and more apt to contain a wood stove than at any other time in history.

The increasing trend is toward smaller prefabricated and mobile homes, which contain synthetic materials more likely to release volatile chemicals to the indoor air.

SUMMARY POINTS (continued)

Certain specific chemicals have been found in levels in the home that were much higher than those outside, but still less than the TLV allowed in the workplace.

Within the past two decades a nonspecific but probably chemically induced constellation of symptoms has come to be known as the sick building syndrome.

The health significance of this source of exposure is not at all clear, but it raises a host of complex policy issues because the air in private homes is not in the public domain.

The regulation of outdoor air quality does not take into account what may be far more intensive exposures to the same chemicals indoors.

Some of the specific chemicals of concern are formaldehyde, asbestos, cigarette smoke, penta, carbon monoxide, oxides of nitrogen, and sulfur dioxide.

BIBLIOGRAPHY

Nero AV Jr: Controlling indoor air pollution. Sci Am 256:42, 1988.

Spengler, JD, Sexton K: Indoor air pollution: a public health perspective. Science 221:9, 1983.

17

CHAPTER

FOOD ADDITIVES AND CHILDHOOD BEHAVIOR DISORDERS

Paul V. Kaplita

THE CLAIM: THE FEINGOLD HYPOTHESIS

Synthetic chemicals have been added to foods for over a century solely to enhance the food's physical appearance. The apparent lack of human toxicity from artificial food colors and dyes is presumed to result from a combination of low exposure levels, incomplete absorption, extensive biotransformation, and complete biliary and renal excretion. Only recently have rational risk-benefit assessments been considered for food colors.

In the mid 1970s, however, the late Dr. Benjamin Feingold, a respected clinical allergist in the San Francisco area, contended that half of all children diagnosed with childhood hyperactivity and learning disorder (minimal brain dysfunction, attention deficit disorder) show a more normal behavior when placed on a special diet. This diet eliminated all artificial colors and flavors, some artificial preservatives, and all foods containing "natural salicylate radicals." He also believed that if children went off his diet it would result in a recurrence of the hyperactive symptoms within hours, and that this relapse might last for several days even if the diet was resumed. He claimed that the diet was so effective that most children on it could forego their usual medications.

The original elimination diet was based on Dr. Feingold's clinical experience with adult patients who had allergy-like reactions to aspirin, one of the so-called salicylate drugs. When the subjects' conditions failed to improve after eliminating all exposure to aspirin and aspirin-like drugs, they were placed on a salicylate-free diet, which also eliminated those salicylates that Feingold thought might occur naturally in foods. The artificial color FD&C Yellow No. 5 (tartrazine) was also eliminated from the diet because it sometimes produced the same allergy-like symptoms in aspirin-sensitive patients.

Because some of his patients did not respond favorably to the diet, Dr. Feingold expanded it to exclude all artificial food additives, even if they were chemically unrelated to aspirin. He noted in anecdotal reports that this diet not only alleviated allergy-like reactions to aspirin in susceptible patients, but it also seemed to improve the behavioral symptoms of some psychologically disturbed patients. So, he began to use the diet to treat children with various behavioral disturbances, including hyperactivity. He eventually justified the diet as a rational treatment by pointing out what he believed was a causal relationship over the last several decades between an increasing use of artificial food additives and an increasing number of clinical reports of childhood hyperactivity and learning disorder.

The original hypothesis and diet underwent many subsequent revisions. Most of the vegetables and fruits that Dr. Feingold identified as containing "natural salicylates," e.g., apples, apricots, cherries, grapes and raisins, oranges, peaches, plums, cucumbers, and tomatoes, actually have only trivial levels of salicylate in comparison with the usual dose of aspirin. Most of the thousands of synthetic chemicals contained in artificial food flavors are, in fact, chemically identical to those that occur naturally in foods; they are simply made synthetically in order to create larger amounts. A subsequent modified diet excluded only artificial colors and flavors not known to occur naturally in food. Most Feingold diets that have been studied in clinical trials do not prohibit artificial flavors because of the difficulty in disguising flavors in the control diets. Moreover, the isolation of any suspected offending compound from the many thousands of flavor constituents in use would be enormously difficult, whereas there were only nine artificial colors in use at the time (below).

Dr. Feingold subsequently altered his original claim to admit that some cases of childhood hyperactivity and learning disorders might be explained on the basis of inheritance instead of allergy. He thought that perhaps some reactions to food colors represented behavioral responses in genetically susceptible children. Later versions of the Feingold diet made by others, excluded the artificial preservatives butylated hydroxyanisole (BHA) and butylated hydroxytoluene (BHT) despite the fact that these ingredients have nothing to do with aspirin sensitivity.

Uncontrolled tests of the Feingold diet seemed to show that it was effective. Feingold Associations, with some 20,000 member families, sprang up across the country. Dr. Feingold recommended legislation that would require a complete ingredient statement on all food product labels, and a symbol on the label to indicate that no synthetic colors or flavors had been added. He also recommended that federally subsidized school lunch programs exclude additive-containing foods. All of his test reports, however, were based on anecdotal observations with no controls and no quantitative assessment of behavior. Despite these gaps in logic and scientific rigor, many became convinced that the diet actually worked, and the food industry, pediatric psychologists, and toxicologists believed that it was worth further testing.

THE FACTS

Childhood hyperactivity or hyperkinesis, the behavioral response hypothesized by Dr. Feingold to result from a hypersensitivity to food additives, has been studied for over 50 years, and almost everything about it is still controversial. The condition is best described as a cluster of signs and symptoms. In the U.S., the syndrome has been reported in 3 to 10% of the child population, with 5 to 10 times greater incidence in males than in females. These rates are much higher than in most other countries. This apparent high incidence might simply be a reflection of differences in medical and social perceptions, instead of actual occurrence. Children diagnosed as hyperkinetic usually show: (1) elevated levels of motor activity, (2) poor control of impulsive behavior, (3) restlessness, (4) short attention span, (5) low tolerance for frustration, (6) irritability, (7) excitability, and (8) aggressiveness. Not every child, however, exhibits all signs and symptoms. Thus, it is even difficult to determine if childhood hyperkinesis is a discrete illness, or a catch-all syndrome with several biochemical and physiological causes. Hyperkinesis could be genetically determined variations in behavioral development, psychogenic disorders related to childrearing, or brain injury caused by trauma, disease, allergy, or environmental toxicity. Some have even suggested that childhood hyperkinesis results from ingesting refined sugar, wearing tight underwear, or prolonged exposure to fluorescent lighting. Stimulant medications such as methylphenidate or *d*-amphetamine produce a "paradoxical" and poorly understood calming effect, but much controversy exists over whether stimulant drugs actually improve learning, behavior, and visual-motor coordination in hyperkinetic children. Other treatments include behavioral modification and counseling, but these therapies have many limitations, involve a long-term commitment, and are expensive.

True food allergies and sensitivities do occur in susceptible individuals. A so-called immediate food allergy produces a reaction that is mediated by an antibody, and it can be detected by skin testing. A delayed food allergy results in diarrhea, vomiting, abdominal pain, dermatitis, or wheezing. It is mediated primarily by special white blood cells, and it often gives ambiguous results on skin testing. Food sensitivity, on the other hand, manifests itself through vague and variable symptoms including aches and pains, fatigue and behavioral disturbances. These symptoms are thought to be the consequence of unusual biochemical reactions with chemical constituents of food when consumed by a genetically susceptible individual.

Dr. Feingold maintained that similar reactions may be produced by the ingestion of food additives. Thus, the hypothesis and diet were an attractive option for families of behaviorally disturbed children, since: (1) the diet offered an alternative to potent stimulant and sometimes abused drugs, (2) the hypothesis relieved parental guilt by transferring the blame from the parents to the food industry, (3) the hypothesis removed any blame from the schools, (4) the hypothesis and diet were consistent with the growing societal awareness at the time about "natural" foods and ecology in general, and (5) the hypothesis empowered parents to control the child's diet so that they in effect become the principal modalities of the treatment.

CLINICAL STUDIES

A controlled, objective evaluation of the effects of food colors on behavior in children was finally designed. It required that homogeneous groups of children be placed, double-blind, i.e., unknown to both observers and subjects, on either a Feingold diet or a control diet and then later to be challenged, double-blind, with either food colors or a placebo. Each child and his family would be interviewed, and the child would be examined for neurologic functions, behavior, achievement, intelligence, learning, and development by a child psychologist and a clinical psychiatrist. The test population would include both normal and behaviorally disturbed children. Behavioral evaluations, involving a combination of global behavioral ratings by parents and teachers, and specific cognitive and learning tests by trained observers would be performed for a baseline period before the study, during the course of the diets, and during the course of the challenges.

To maximize full participation in the trial, tasty nutritious diets would be administered either in a hospital inpatient setting or at home with the entire family participating. The most appropriate food color challenge was thought to be a blend of the nine colors approved for use in the U.S. and Canada (3.12% FD&C Blue No.

1, 1.70% FD&C Blue No. 2, 0.13% FD&C Green No. 3, 6.08% FD&C Red No. 3, 0.50% FD&C Red No. 4, 38.28% FD&C Red No. 2 or No. 40, 26.91% FD&C Yellow No. 5, 22.74% FD&C Yellow No. 6 and 0.54% Orange B). The proportions of each individual color in the blend were based on the amount actually used on a per capita basis in food production.

However, there were several problems with this study design, which eventually obscured the final results. It was extremely expensive, and many of the clinical studies were supported by the Nutrition Foundation, which is in turn supported by the food industry. This led to charges of subjective interpretation of the data. Often, the validity of particular psychological and behavioral test methods were questioned. There was little attention paid to dose-response relationships in the studies. Finally, the chocolate that was used to mask the presence of food colors in the challenge in many studies was criticized, because chocolate was thought by some to elicit behavioral effects of its own.

The earliest tests of Feingold's hypothesis involved simply placing children diagnosed as hyperkinetic on a Feingold diet and observing any changes in behavior. There was certainly no lack of tests. One such study reported that ten of fifteen hyperkinetic children showed improved behavior while on the diet, according to questionnaires filled out by their parents. Another noted that parents, teachers, and pediatricians observed improvements in behavior in 11 of 32 hyperkinetic children placed on the diet. Yet a third tested the allergic responses of 31 hyperkinetic children to four food colors by monitoring blood pressure, pulse, and the incidences of headache, rash, pallor, perspiration, and activity. Eighteen children reacted to the colors, and 15 of these children were placed on a Feingold diet for four weeks. According to parental rating scores, the children's behavior improved by 93% while on the diet. In another report it was noted that the behavior of two hyperkinetic girls deteriorated significantly following the ingestion of a mix of artificial food colors. Others reported that a hyperkinetic boy responded positively to a Feingold diet, as observed by parents and teachers. These anecdotal reports, of course, were no more scientifically sound than Dr. Feingold's original observations.

Several controlled studies were performed which refuted Feingold's hypothesis. In one of these 15 children evaluated as hyperkinetic were observed by parents and teachers over a 4-week baseline period. The children then received a Feingold diet and a control diet, each for four weeks, in a double-blind, random, counterbalanced crossover design, in which half the children received first one diet and then the other, and the other half received the diets in the opposite order. With this design, each child served as its own control, and any effect produced by the diet resulting from expectations or suggestibility of the child, parents, or other observers can be detected. The control diet contained food colors, but it was not a true control because it was explained to the subjects that it was an alternate elimination diet involving other food constituents.

Both test diets were nutritionally equivalent to a normal diet, but the Feingold diet contained slightly less vitamin C and carbohydrates than normal. Teachers concluded that there was an improvement in behavior in children on the Feingold diet relative to the baseline and control diet periods using behavioral rating scales. The differences between the effects produced during the Feingold and control diets, however, were observed only when the control diet preceded the Feingold diet. Parents detected improvements in behavior only between the baseline and Feingold diet periods, and again only when the control diet preceded the Feingold diet. These results are typical of an "order effect" produced when the effects of a treatment combine with a factor of suggestibility to amplify the effect of treatment when it follows a control period.

In a second study, sixteen children who showed a significant response to the Feingold diet were then observed by parents and teachers over a 1-week baseline and a 3-week diet period. During 8 additional weeks on the Feingold diet, the children received twice daily challenges of food color or placebo (inactive control material), alternating every 2 weeks in a double-blind random, counterbalanced crossover design. The active challenge consisted of chocolate cookies that contained 13 mg of the food color blend. The placebos contained no color blend. Neither parents nor teachers detected any behavioral effects from the challenges, as measured by the ratings scales. In a tracking task that required the child to keep a spot of light on a moving target while having to respond to intermittent flashing lights on either side of the tracking task, however, the children performed less well within the first hour of consuming the active challenge than after placebo or in control tests.

To determine whether the results of the tracking task represented a short-term effect of a single large dose of food colors, 13 children who showed a significant response to the Feingold diet were observed by parents and teachers for 3 hours following challenge with either food colors or placebo. After observing the children's behavior for a 2-week baseline period and for 3 weeks on the diet, the parents evaluated their children's behavior immediately following twice daily challenges with cookies containing either 13 mg color blend or placebo. After one week the challenge was reversed. According to rating scales, 4 of the 13 children showed a significant worsening of behavior immediately after the color challenge.

The above study was then repeated, but with thirty children who after initially responding to the Feingold diet, received either color challenge or placebo twice daily, alternating each week for 4 weeks in a double-blind, random, counterbalanced crossover design. The parental ratings scores revealed no differences in any child's behavior for the 3 hours following either active challenge or placebo. The challenge study was repeated once more, with behavior and restlessness evaluated by trained observers, activity levels measured by a counter, and paired-associated learning tasks performed before and during the four hours after either the color challenge or placebo. Nine children who responded favorably to the Feingold diet later received twice daily challenges or placebos in a double-blind, counterbalanced crossover protocol. None of these children showed any differences in any of the behavioral evaluations before or after the color challenge or placebo.

Similar studies also reported subtle, inconsistent results with a Feingold diet. After being observed for a 2-week baseline period, 46 hyperkinetic children and their families received 4 weeks of the Feingold diet and 4 weeks of a control diet in a double-blind crossover protocol. This particular control diet was designed in such a way as to make it indistinguishable from, yet nutritionally equivalent to, the Feingold diet. The teachers and trained observers detected no differences in children's behavior during either the diet or baseline periods. The parents did notice an improvement in behavior during the Feingold diet, but again only when the control diet preceded the Feingold diet.

The 9 children who showed the most significant changes while on the Feingold diet above were observed for an additional 2 weeks of baseline and 2 weeks of diet. Then, over a 9-week period they received twice daily challenges of food colors or placebo, alternating every 2 weeks in a double-blind crossover design. The active challenge again consisted of cookies that contained 13 mg of the color blend. No significant differences were seen in the behavioral ratings of parents, teachers or observers between periods of active challenge or placebo, although one child did show some slight response to the colors.

Because the test results to date were less than satisfactory, a decision was made to examine the effects of larger doses of food colors on children's behavior. A hyperkinetic boy was placed on the Feingold diet and subsequently challenged twice weekly over a 10-week period with either 39 mg of the color blend or a placebo in a double-blind design. Although teacher and child could detect no behavioral changes following challenge or placebo, the child's mother often noted a deterioration of behavior following the color challenge. This experiment was extended to examine the effects of 3 daily challenges with either 78 mg of the color blend or placebo, administered to 13 hyperkinetic children in a double-blind crossover design. No differences were observed between reactions following color challenge or placebo in parental or teacher behavioral evaluations, or psychiatric tests.

In another study, 22 children, placed on a Feingold diet for at least 3 months, received random, double-blind challenges over a 7-week period in the form of soft drinks containing either 35.6 mg of a blend of food colors or placebo. The amount of FD&C colors in this blend represented the amounts of colors consumed daily by the general population. Twenty of the children showed no reaction to the challenge, but 2 (1 of which was the smallest of the group) reacted after each color challenge with disturbed behavior, as measured by parents and teachers using a ratings scale.

Others examined a select sample of 40 children with behavioral symptoms suggesting childhood hyperkinesis. Twenty of the children had previously responded favorably to stimulant medication but scored poorly on behavioral ratings scales. These were considered "confirmed" hyperactives. The remaining 20 showed adverse reactions to stimulant therapy and scored high on behavioral ratings scales; their diagnosis was "unconfirmed." The basis of these diagnoses, however, has been questioned. After 5 days on the Feingold diet as hospital inpatients, 20 of the children (10 from each diagnosis group) received a capsule containing either 100 mg of the food color blend or placebo. The other 20 children received a capsule containing either 150 mg color blend or placebo. This single dose of the color blend falls within the 90th percentile for daily consumption of artificial food colors by 5 to 12 year old children.

After the administration of the color blend, those children with confirmed hyperkinesis showed a significant impairment in paired-associated learning performance relative to their performance before dosing. Those children with an unconfirmed diagnosis showed no signficant differences in learning performance before or after dosing. The reaction in the confirmed group took over a half hour to become apparent, and it reached a maximum intensity after an hour and a half. There was no difference between the behavioral ratings of either group after color challenge or placebo. The interpretation that these results demonstrate a short-term effect on cognitive function has been questioned, however, because the responses of the confirmed group after placebo differed from the responses of the unconfirmed group after placebo.

In additional studies the relative effects of a Feingold diet versus a control diet that contained food additives on 39 children diagnosed with moderate-to-severe learning disabilities were examined. Eighteen of these children were regarded as hyperkinetic, and 17 were being treated with various medications during the course of the study. In a double-blind design, the children were observed by teachers during 1 week on a Feingold diet, followed by 1 week on the control diet. No significant behavioral differences were detected by the teachers.

Two studies specifically examined the effect of FD&C Yellow No. 5 (tartrazine) on children with behavioral problems. After a period of baseline observations, 30 children were placed on a Feingold diet for 4 weeks. The children were then challenged daily over the next 2 weeks with either 4 to 5 mg tartrazine or placebo in a double-blind crossover design. The children then continued on the Feingold diet for another 4 weeks. Teacher evaluations and cognitive and learning tests detected no effect on the children's behavior by the diet, the color challenge, or the placebo. Parents, however, noticed a significant improvement in behavior, but only during the first 4 weeks on the Feingold diet. Thirteen children who showed the most favorable responses to the diet underwent a second challenge period. Behavioral reactions to the tartrazine challenge were detected by parents, but not by the teachers and not by the cognitive and learning tests. The 8 children that did react to tartrazine received a third challenge trial. This time, the parental evaluations detected no differences in behavior after either tartrazine or placebo.

The effects of a Feingold diet, and subsequent challenge with tartrazine and the red food color, carmoisine, on children referred for suspected hyperactivity were also examined. Of 220 children originally referred, 55 were enrolled in a 6-week open trial of the Feingold diet. Eight of these children responded to the diet and were then enrolled in a double-blind crossover protocol, employing a single-subject repeated measures design. The children were challenged daily with either 50 mg tartrazine or carmoisine, or placebo for 18 weeks. Two children reacted to the color challenge with increased irritability and restlessness and disturbed sleep patterns.

The relative therapeutic effects of stimulant medication and a Feingold diet on 26 hyperkinetic children were also studied. After 1 week of baseline observations and 5 weeks on the Feingold diet, the children received either stimulant medication or placebo, and 4 challenges per week in the form of cookies containing 26 mg color blend or placebo, in a double-blind, random, counterbalanced crossover design. Behavioral responses were measured using parental and teacher evaluations. Stimulant medication produced significant improvement in the children's behavior. Neither the active challenge nor the placebo had much effect while the children were on the stimulant medication. The effects of the diet were inconsistent, with parents noting no changes in behavior during the time the children were on the diet or undergoing challenges, whereas teachers observed that the children showed the worst behavior following an active challenge while on placebo medication. Since all challenges were administered during school hours, however, the behavioral deterioration observed by the teachers could have been a short-term reaction that disappeared by the time the children left school at the end of the day.

Thus, Feingold diets generally do not appear to work in controlled, blind studies. A National Institute of Health Consensus Development Conference concluded that "the *Feingold* diet *may* be helpful for a *small* number of children with hyperkinesis." They estimated that number to be anywhere from 25 to 250 per 100,000 patients, but they also noted that these decreases in hyperactivity were not observed consistently. Some physicians believe that the isolated favorable responses that have been reported justify yet further extension of clinical studies, but others are ready to quit. The best responders to the diet and challenges were often the smallest, youngest children. Although this observation may reflect a development influence, the magnitude of responses to a consistent amount of color challenge in different children could be a function of body weight, with smaller children actually receiving a larger mg/kg dose than older, larger children. Dose-response relationships could be examined for color blends and individual FD&C dyes, perhaps using capsules instead of cookies as the challenge vehicle to minimize jeopardizing the placebo disguise at higher doses. The absorption, distribution and elimination of the dyes, including whether or not they cross the blood-brain barrier, could also be studied. It might be of interest to alter the frequency of dosing with the challenges, in order to more accurately represent real-life daily consumption of food colors.

ANIMAL EXPERIMENTS

Studies with rats suggested that FD&C Red No. 3 (erythrosin B) might be the most active agent in food color blends. Rat pups can be treated with a drug that produces a childhood hyperkinesis-like syndrome, including persistent learning deficits. In pups so pretreated 0.5 to 2.0 mg/kg color blend daily by mouth showed increases in motor activity relative to controls that received the drug alone although no consistent dose-response relationship was observed. Moreover, the colors had no effect on the pups' learning performance in the T-maze and shuttle-box. However, pups pretreated with water and then administered 2 mg/kg color blend also showed increases in activity, and exhibited impaired learning performance when compared with controls. Similar effects on activity and learning performance were seen in naive and drug-pretreated rat pups in which 1 mg/kg/day color blend had been placed in their stomachs. Daily oral administration of 1 mg/kg erythrosin B, however, failed to elicit any significant effect on the motor activity or cognitive performance of developing rat pups. Likewise, erythrosin B fed at up to 1% of the diet produced no consistent, dose-dependent behavioral effects or developmental toxicity in pregnant rats, rat dams, or rat pups.

In another study, injected erythrosin B at 50 mg/kg had no effect on motor activity in rats pretreated with drug or a vehicle. Rats treated with 50 to 300 mg/kg erythrosin B, however, showed significantly increased punished-response behavior relative to controls. This effect was dose-related and was similar to the change in punished-response behavior seen following such drugs as barbiturates. Interestingly, barbiturates also worsen the symptoms of childhood hyperkinesis in humans, and this effect can be blocked by stimulant drugs.

Erythrosin B is a commonly used synthetic dye approved by the Food and Drug Administration for use in foods, drugs, and cosmetics. General toxicology studies indicated that erythrosin B is relatively nontoxic. Erythrosin B administered orally to beagles at 0.5 to 5.0% of dietary levels for two years produced no apparent toxic effects, but in rats it caused intestinal distension, reduced spleen weight and depressed growth at the highest dose that was administered. An acute LD50 for oral administration in rats was calculated as between 2 and 7 g/kg. At toxic doses, the animals' internal organs, muscle, fat, ears, tails, paws, and noses took on a pink coloring from the absorbed dye. Pigs treated for ninety days with erythrosin B, showed alterations in thyroid weight and serum thyroid hormone levels, but this did not occur in rats treated for eighty-five weeks. High doses of erythrosin B showed slight, but statistically significant mutagenic effects. The acceptable daily intake of erythrosin B was established as 2.5 mg/kg. The average daily dietary intake of erythrosin B has been estimated at 34.1 mg, which for a 70 kg human is equivalent to a daily exposure of 0.5 mg/kg. In rats, oral doses of erythrosin B are largely excreted in the feces although a small amount is also secreted in bile. The distribution and metabolism of oral doses of erythrosin B in humans are not known, although it is thought that humans also excrete the dye largely in the feces. Any correlation between behavioral studies and biochemical and physiologic experiments must be made cautiously, because in rats erythrosin B does not penetrate the blood-brain barrier to enter the central nervous system after systemic administration.

Interestingly, erythrosin B and analogs exerted toxic effects on several species of insects, including house flies, boll weevils, and mosquitoes. Dietary levels of 0.5 to 5.0 mM erythrosin B produced nearly 100% lethality in sample insect populations. This corresponded to low whole body concentrations in the insects. The insecticidal effect depended on both dose and duration of exposure. At least two mechanisms of action seem to be involved: a rapid, dye-sensitized photo-oxidation that uses energy from trapped photons of light to transform molecular oxygen into toxic free radicals, and a slower, light-independent reaction. The dark reaction occurred at higher doses of the dye than those producing the light-dependent toxicity.

Erythrosin B also rapidly and reversibly inhibited fertilization in sea urchin eggs. This effect depended on concentration, and it occurred in the absence of any change in sperm or egg viability. Other analogs of erythrosin B also inhibited fertilization. Their relative activities, measured as the concentration that inhibited fertilization in 50% of a sample of eggs, increased in direct proportion to the degree of fat solubility of the compounds.

Erythrosin B and several analogs also produced membrane effects on neurons in isolated nerves of a mollusk. Curiously, after exposure to the dyes, the connective tissue surrounding the nerves remained stained, even after extensive washing. Although the immediate membrane effects were reversible, it was noted that exposure to eosin Y, erythrosin B, phloxin B, and rose bengal also initiated a slow, irreversible decline in the resting membrane potential. Effects have also been described on the frog neuromuscular junction, rat renal cortex, and neurotransmitters in rat brain. The concentrations of erythrosin B studied also stained the tissue homogenates red.

SUMMARY POINTS

When fed under controlled conditions, Feingold elimination produced subjective and equivocal changes in the behavior of disturbed children; the subject was heavily studied because all parties wanted so badly for it to be true.

Few instances of consistent, dramatic deterioration in behavior were seen after double-blind challenge with artificial colors during the period of diet enrollments.

The major remedial effect of the diet appears to result from a placebo response arising from changes in the attitudes of parents and children about the condition, positive expectations regarding the diet, and an underlying belief that certain components of processed foods are toxic.

Over the course of the diet period, the child receives much more attention from family members, and any blame or guilt concerning the child's condition is deflected to the family's diet; thus, the diet acts almost as a form of behavior modification.

SUMMARY POINTS (continued)

The biochemical studies, however, offer evidence that FD&C Red No. 3 may be neurotoxic.

The behavioral effects observed in the animal studies and the limited number of responders in the clinical studies suggested that a small percentage of hyperkinetic and learning disabled children may be adversely affected by one or more food colors.

Feingold diets are safe as well as nutritionally adequate, and the National Institutes of Health Consensus Committee has advised that affected children may try the diet as long as other therapeutic options, like medication, are not discarded.

Although Feingold's legislative recommendations were not widely implemented, he served to focus public attention on food additives and food package labeling.

BIBLIOGRAPHY

Brenner A: A study of the efficacy of the Feingold diet on hyperkinetic children. Clin Pediatr 16:652, 1977.

Burlton-Bennet JA, Robinson VMJ: A single subject evaluation of the K-P diet for hyperkinesis. J Learn Disabil 20:331, 1987.

Conners KC: Food Additives and Hyperactive Children. New York, Plenum Press, 1980.

Feingold BF: Hyperkinesis and learning disabilities linked to artificial food flavors and colors. Am J Nursing 75:797, 1975.

Feingold, BF: Why Your Child is Hyperactive. New York, Random House, 1975.

Kaplita PV, Triggle DJ: Food dyes: behavioural and neurochemical actions. Trends Pharmacol Sci 3:70, 1982.

18

CHAPTER

HAZARDOUS WASTE SITES

Dennis J. Devlin

The volume of chemical waste produced in industrialized nations including the U.S. is astonishing. Nearly 6 billion tons have been produced in the U.S. since 1950, and approximately 300 million tons of waste are now produced annually. To put this in some perspective, that annual figure is equivalent to more than 1 ton per U.S. citizen per year. It is a volume of waste that would fill the New Orleans Superdome 1500 times. These figures are alarming because much of the waste contains chemicals that are potentially hazardous to human health and the environment. Waste production has been going on for over a century, but it is only within the last decade that the enormous and complex problems associated with waste disposal have been addressed. Efforts are at last underway to reduce the hazards inherent in the storage, transportation, and disposal of wastes currently being produced through government regulation and recycling of waste into useful products. However, additional effort is needed because more than 10,000 uncontrolled waste sites created in the past now present an imminent danger. Many of these sites are abandoned, intentionally hidden, or simply forgotten. The understanding and assessment of the hazards associated with industrial waste disposal is an important practical application of environmental toxicology.

LOVE CANAL—THE ALARM SOUNDS

Among the thousands of known industrial waste sites in the U.S., one will be remembered for awakening the population to the potential risks and community anguish associated with mismanaged disposal of hazardous waste: that site is Love Canal, in Niagara Falls, NY. Although those responsible for this site have been criticized for an appalling lack of foresight, for that point in time, state-of-the-art waste disposal practices actually were followed. A review of the history of Love Canal from its innocent beginnings to its tragic impact on the local community is a general chronicle of the history of many abandoned waste sites.

In the early 1900s, William Love began building a 7-mile canal to connect the upper and lower levels of the Niagara River, but the project was abandoned after only 3000 feet had been excavated. Thirty years later, Hooker Electrochemical Corporation was granted permission to dispose of its waste in the half-mile dig. Over the next 11 years more than 21,000 tons of liquid and solid waste, primarily from organic pesticide production, were dumped in the canal and covered with what was then believed to be an impermeable layer of clay. In 1953, the land was sold to the Niagara Falls Board of Education for $1. This land and the surrounding area were subsequently developed for residential use. An elementary school was built over the site. The physical integrity of the impermeable clay cap covering the waste was breached during the construction of roads and buildings. The critical and perhaps inevitable event had occurred. Melting snow and rain percolated through the canal saturating the waste, and collecting soluble contaminants in the process. This toxic leachate later entered the culverts, backyards, and basements of the nearby homes. Tests on air, water, and soil in the area revealed the presence of over 80 hazardous chemicals, including several suspected carcinogens. Lindane, chloroform, benzene, and dioxin were among the toxic chemicals found. In 1978 over 200 families were evacuated from the area, and nearly 1000 families were included in a federal declaration of a health emergency. To the frustration of many residents, the scientific and medical communities were unable to reach a consensus on the health-related consequences of exposure to the canal leachate. The extent of actual illness caused by the leachate will be debated for years, but there is little doubt that some residents suffered from the uncertainty concerning their health and from the permanent dissolution of their community.

PAST WASTE DISPOSAL PRACTICES

Love Canal was not an isolated example of haphazard waste disposal. Waste disposal policy beginning with the Industrial Revolution and continuing through the mid 1970s was founded on the canard that the earth could assimilate or transform hazardous waste into harmless products or dilute it to safe concentrations. This policy was fueled by avarice to minimize the costs associated with waste management. Consequently, waste was disposed of at locations that were both convenient and inexpensive, ergo, disposal at or near the site of production. The method of disposal was contingent upon the form of the waste: liquid, solid, semisolid, or gaseous. Solid and liquid wastes were sometimes incinerated. Liquid wastes were often placed in surface impoundments such as lagoons or ponds. Although some impoundments were lined with relatively impermeable materials such as clay, most were constructed over porous or sandy soil to allow for seepage into the underlying earth. This design permitted continual replacement of liquid waste into the impoundment.

Some liquids were indiscriminately discharged onto the land or into the surface water of rivers and lakes. Liquids and solids were often placed into 55 gallon steel drums or larger tanks at the factory and then buried in unsecured landfills or existing mines. Some solid wastes were merely dumped into piles and left uncovered. Worst of all, some hazardous wastes were combined with domestic waste in community landfills. Many of these landfills were not designed to prevent the migration of hazardous constituents. The enormous size of these landfills may make abatement of this migration a practical impossibility. Although intended to be forgotten, many of these ill conceived waste disposal sites are now leaking contaminants, and raising concerns about human health and the environment. What once seemed an economically prudent policy of dumping useless waste has created one of our primary environmental crises. Its resolution could consume $35 billion.

HAZARDOUS WASTE SUPERFUND

The public attention given Love Canal galvanized federal, state, and local governments, and citizen groups into organizing searches for industrial waste that had been dumped and abandoned. The term, toxic waste dump, was coined to describe these locations. This extensive search also identified sites that were potentially dangerous where the haphazard handling and storage of usable chemicals was leading to an imminent threat of environmental contamination. Many of the substances identified were hazardous for reasons other than their toxicity (see Chapter 2). Some chemicals were explosive, flammable, or highly reactive if mixed with air or water. Thus, the more appropriate term, hazardous waste site, evolved to describe these locations. As for the term, waste, it technically and legally refers not only to materials that have served their intended purpose and have been discarded, but also to usable chemicals that have escaped their containers and entered the environment.

By 1980, thousands of hazardous waste sites had been identified. Many were releasing hazardous substances into the environment via groundwater, surface water, and air. However, it was often difficult to identify those responsible, and of those who were identified, few had the financial means to abate the hazard. Others refused to accept liability for the hazard because their disposal practices had been acceptable at the time. Congress recognized the need for a nationally coordinated effort to fund cleanup of these inactive hazardous waste sites, and they passed the Comprehensive Environmental Response, Compensation, and Liability Act of 1980 (CERCLA). The Act is best known for the "superfund" it created. CERCLA provided for the accumulation of $1.6 billion between 1980 and 1985, which came primarily from a tax on industrial chemicals. The fund can be used for the following: emergency cleanups, removing waste from the site for safer disposal, or remedial environmental action such as decontamination of soil and groundwater. The fund can also be used to compensate individuals for loss of property or damage to property resulting from an abandoned hazardous waste site.

The fund cannot be used to compensate for damages to human health or death resulting from exposure to the waste. The original act of 1980 contained a provision by which the law was to be reviewed, amended, or abandoned after 5 years. In 1986, the Superfund Amendments and Reauthorization Act (SARA) was enacted to extend the superfund, totaling $8.5 billion over the next 5 years. In 1990, Congress again reauthorized CERCLA, allowing for an additional tax of up to $5 billion to supplement the superfund between 1991 and 1995.

THE NATIONAL PRIORITY LIST OF WASTE SITES

An important provision of CERCLA was the establishment of a national priority list (NPL). The NPL identifies those sites that are believed to present the greatest danger. By 1990, over 1000 sites were proposed for the NPL. None of the 50 states is without a site on the NPL. Not surprisingly, the states with the most priority sites are known primarily for their industrial economies, whereas those with the fewest are less industrialized. In keeping with the philosophy noted above, industrial waste was not transported far from its site of production irrespective of the suitability of the local environment for disposal.

Investigations at these waste sites identified some of the chemicals that were hazardous. Most sites contained more than one

TABLE 18–1. STATES REPORTING THE MOST NATIONAL PRIORITY LIST SITES

State	Total	State	Total
New Jersey	109	Texas	28
Pennsylvania	95	Massachusetts	25
California	88	South Carolina	23
New York	83	Missouri	22
Michigan	77	North Carolina	22
Florida	51	Delaware	20
Washington	45	Iowa	20
Minnesota	42	Virginia	20
Wisconsin	39	Kentucky	17
Illinois	36	Colorado	16
Indiana	33	New Hampshire	16
Ohio	33	Connecticut	15

and some contained hundreds. Many sites contained the same chemicals, namely, those found in common industrial waste streams. For example, similar solvents and degreasers were used in many manufacturing industries, including several halogenated organic compounds, such as chlorinated ethanes and ethylenes (two cojoined carbons with attached chlorine atoms; see trichloroethylene below). Once the solvent or degreaser was contaminated by use, it was discarded.

THE PRIORITY POLLUTANTS

The great majority of chemicals found at hazardous waste sites are "priority pollutants." Priority pollutants are a group of 129 chemicals previously designated pursuant to the Clean Water Act of 1977. These are believed to pose the greatest and most prevalent hazard to aquatic environments. The priority pollutants can be divided into nine subgroups of chemicals. These subgroups and some common examples are found in Table 18–2. The subgroups most often found at hazardous waste sites are the halogenated aliphatic hydrocarbons (such as those listed above), monocyclic aromatic hydrocarbons (containing one benzene ring), and metals (see Chapter 12). Priority pollutants that do not fall into any of the nine subgroups listed include cyanide, asbestos, and benzidine. Furthermore, not all hazardous materials are identified as priority pollutants. Hazardous chemicals not listed as priority pollutants include radioactive substances.

TABLE 18–2. THE PRIORITY POLLUTANTS

Class	Examples
Metals and inorganics	Lead, cadmium, cyanide
Pesticides, PCBs, related compounds	Aldrin, DDT, TCDD, PCB
Halogenated aliphatic hydrocarbons	Trichloroethylene, bromomethane
Halogenated ethers	Bis(2-chloroethyl) ether
Monocyclic aromatics	Benzene, toluene, phenol
Polycyclic aromatic hydrocarbons	Benzo[a]pyrene, naphthalane
Phthalate esters	Di(2-ethylhexyl) phthalate
Nitrosamines	Dimethyl nitrosamine
Miscellaneous compounds	Benzidine, acrylonitrile

The Agency for Toxic Substances and Disease Registry (ATSDR) has been mandated by Congress to produce toxicologic profiles for hazardous substances found at NPL sites. These substances were ranked on the basis of the frequency of their occurrence at these sites, their toxicity, and the potential for human exposure. The first 200 substances were ranked by the end of 1988. The 25 highest ranked substances are listed in Table 18–3. The ten chemicals most often found at sites on the NPL are described below. Since they have been released into the environment from many sites, some knowledge of their use and toxicity is important for those interested in environmental toxicology.

Trichloroethylene

Trichloroethylene (TCE, $CCl_2{=}CHCl$) is a clear colorless liquid at room temperature. It is used extensively as an industrial degreasing solvent, chemical intermediate, and heat transfer agent. Since

TABLE 18–3. THE HIGHEST RANKED SUBSTANCES BY ATSDR OF THOSE FOUND AT NPL SITES

Aldrin/Dieldrin	1,4-Dichlorobenzene
Arsenic	Di(2-ethylhexyl)phthalate
Benzene	Heptachlor/Heptachlor epoxide
Benzo[a]anthracene	Lead
Benzo[a]pyrene	Methylene chloride
Benzo[b]fluoranthene	Nickel
Beryllium	N-Nitrosodiphenylamine
Cadmium	Polychlorinated biphenyls (PCB)
Chloroform	2,3,7,8,-Tetrachlorodibenzo-p-dioxin (dioxin)
Chromium	Tetrachloroethylene
Chrysene	Trichloroethylene
Cyanide	Vinyl chloride
Dibenzo[a,h]anthracene	

1979 more than 230,000 tons of TCE have been produced annually in the U.S., although its use is decreasing. TCE is readily absorbed after inhalation or ingestion. and it is relatively nontoxic after acute dosing. Exposure to TCE may result in symptoms of central nervous system (CNS) depression with headaches, mental confusion, vertigo, insomnia, and fatigue. Prolonged exposure can produce impairment of the peripheral nervous system as well, up to and including paralysis. A chronic inhalation study of TCE in mice indicated an increased incidence of lung and liver cancer and leukemia.

Lead (See also Chapter 12)

Lead (Pb) is a soft, naturally occurring metal used in a wide variety of industries including electroplating, metallurgy, and plastics. It is also an important component of batteries, electronic equipment, and pigments. Its use as a gasoline additive was once extensive, but it has now been phased out. Lead can be found in an inorganic form as an oxide (lead dioxide, PbO_2), or as a salt (lead chloride, $PbCl_2$), or in organic forms as alkylated lead. The central and peripheral nervous systems, kidneys, reproductive system, and the blood forming hematopoietic system can be adversely affected by lead. Although lead produces kidney tumors in rats, no epidemiologic data exist to suggest that it is carcinogenic in humans. Several studies suggest that lead is embryotoxic, and may have caused miscarriages in women working with it.

Toluene

Toluene or methylbenzene is another widely used industrial solvent. It is used as a component in gasoline and other organic mixtures, and in the manufacturing of paints, glues and enamels. Toluene has CNS depressant properties, and it produces symptoms of fatigue and weakness when inhaled. It is relatively nontoxic to other systems and organs of the body. No evidence suggests that toluene is carcinogenic or teratogenic in humans or animals. Although not considered highly toxic, it is flammable, and it contributes to the fire and explosion hazard at many waste sites.

Benzene

Benzene is a volatile liquid produced as a by-product in the petroleum, coal processing and coke oven industries. It has had many industrial uses, such as in the production of drugs and chemicals, degreasing and cleaning solvents, and rubber and plastics, and as a fuel additive. Over 4 million tons of benzene are produced annually in the U.S. Benzene accumulates in the fat and bone marrow, and its biotransformation products are sequestered by the liver and bone. Acute exposure to benzene produces symptoms of CNS depression leading to loss of consciousness following exposure to high concentrations. A reduction in the number of blood cells and impairment of the immunologic system have been reported in workers exposed chronically to benzene. Evidence exists that long-term exposure to benzene may cause leukemia in humans.

Chloroform

Chloroform or trichloromethane is a pleasant smelling solvent used in the manufacturing of plastics, drugs, and refrigerants. Over 100,000 tons are produced annually. Chloroform is efficiently absorbed from the gut, lungs, or skin. It is sequestered by all organs of the body, but its primary targets are the liver and kidneys. As with the other solvents mentioned above, chloroform also produces CNS depression when inhaled. It has been found experimentally to produce liver tumors in mice and kidney epithelial tumors in rats.

Other Hazardous Chemicals

Of the remaining five chemicals on the list of those most commonly encountered at hazardous waste sites, PCB (see Chapter 11) and cadmium (see Chapter 12) are considered to be carcinogenic. Tetrachloroethylene and 1,1,1-trichloroethane are popular solvents used in a variety of industries. They possess the toxic characteristic of CNS depression after inhalation typical of many solvents. While the above chemicals are among those most commonly found at hazardous waste sites, they are not necessarily the most hazardous chemicals found at or released from these sites. Known or suspected human carcinogens, including asbestos, and hexavalent chromium, and acutely toxic chemicals such as cyanides and sulfides may present an immediate risk at the sites where they are found. Extremely corrosive, flammable or explosive substances also warrant special care.

Besides the primary dangers of fire and explosion, some wastes can produce highly toxic fumes when burned. These fumes may contain phosgene and other pyrolysis products of organic chlorides. Other on-site hazards result from combining incompatible wastes, which might produce fire, heat, explosion, or toxic gases when mixed. Examples of pairs of incompatible wastes include: acids and bases, strong oxidizers such as peroxides or perchlorates and combustible wastes, and alcohols and acids. At one site, rusting drums of sodium cyanide and various acids were found side-by-side. If these contents were to mix, dangerous cyanide gas would be released.

HUMAN CONTACT WITH WASTE CONSTITUENTS

Improperly stored or discarded hazardous wastes can endanger human health through direct contact if the site is not secured, or more commonly through the migration of waste constituents from the site. Surface impoundments, unlined landfills, and rusting drums and tanks eventually release waste chemicals into the environment. Solid wastes are leached by rain or snow that percolate through the site and carry out contaminants. Toxic leachate and liquid waste contaminate groundwater by direct seepage through porous soil and contaminate surface water via surface drainage or flooding of the site. Toxic particulates from sites such as exposed piles of pesticides and asbestos can become airborne creating an inhalation hazard. Gaseous and volatile wastes, including organic solvents, will escape from corroding containers to produce toxic vapors, especially in warmer climates.

Contaminants released into air or surface water are substantially diluted, which tends to reduce their immediate hazard, but dilution is not always enough to totally eliminate the danger from environmental exposure. The most common health complaints of people living or working near a waste site are headache, dizziness, nausea, and fatigue. These common symptoms are associated with the inhalation of hydrocarbon vapors such as kerosene. The primary complaint at the Love Canal site was caused by volatile organic chemicals that entered basements and contaminated the air inside homes. At other sites, a trivial amount of pollution, combined with other sources, produced an additive adverse effect on the surrounding environment.

Human exposure to contaminants released into surface water may occur if the water is used directly for drinking, cooking, or bathing. Exposure may also occur from ingestion of fish or other seafood taken from the contaminated water. As mentioned in Chapter 9, certain contaminants are known to bioaccumulate significantly in the aquatic food chain. The bioconcentration factor

(BCF) was developed to estimate the extent of bioaccumulation. The BCF is defined as the experimentally determined ratio of the concentration of the chemical in the tissues of a specified aquatic organism, divided by the concentration of the chemical in the organism's water environment. Although the entire organism is usually digested to determine the BCF, the concentrations of the chemical usually are higher in fatty tissue and lower in leaner tissue. Examples of contaminants that bioaccumulate significantly and their associated BCFs are methylmercury (6200), DDT (39,000), and PCBs (60,000). Chemicals that do not bioaccumulate significantly include chloroform (14), cyanide (2.3), and arsenic (1).

CONTAMINATION OF GROUNDWATER

Groundwater, surface water, and air are all susceptible to contamination from hazardous waste sites. However groundwater contamination poses the most critical, long-term hazard because it serves as the primary water source for nearly half of the U.S. population. Contaminants can be present in groundwater at levels exceeding standards set to protect human health, and yet go undetected for long periods of time. Detection of groundwater contaminatation is often delayed unless aesthetic qualities such as the taste, odor, or color is affected. Most private wells are not routinely monitored to detect organic contaminants typically found in industrial waste. Detecting these contaminants at the low levels commonly found in groundwater requires more sophisticated and expensive analytical techniques than those routinely used, such as gas chromatography and mass spectroscopy. Once a groundwater source is contaminated beyond safe use, decontaminating it to safe levels is often impossible. In most cases the contamination persists for many years after the surface source is eliminated, and a new water supply must be located.

Groundwater exists as aquifers, highly permeable underground strata, which are analogous to underground rivers. Approximately 97% of all fresh water on earth is located in aquifers. Aquifers are classified as unconfined, or water table aquifers, and confined, or artesian aquifers. Zones of greatly reduced permeability separating aquifers are known as aquitards or confining layers. Unconfined aquifers are located above the shallowest confining layer. The water volume fluctuates seasonably with changes in precipitation and surface freezing. The surface of unconfined aquifers may lie anywhere from ground level to great distances underground depending on the rate of water recharge from precipitation and surface water such as streams. Because the water table has no confining layer above it, it is the aquifer that is most susceptible to contamination by surface pollution. Some private wells draw water from this uppermost aquifer, whereas most municipalities rely on groundwater drawn from deeper artesian sources to ensure greater protection of the water quality. The movement of contaminants between the unconfined aquifer and lower artesian aquifers occurs at natural fractures in the confining layer or through manmade breaks including unprotected or damaged well casings.

Contaminants in the groundwater move by a definite route within the aquifer at flow rates on the order of a few inches to several feet per day, depending on the permeability of the aquifer and the existing hydraulic gradient. The contaminants can be dissolved in the ground water or can move as a separate chemical phase independent of the flow of groundwater. The moving body of contaminants is called a plume, as also is the effluent of airborne materials from a smoke stack. The shape and size of the plume depends on the geology, groundwater flow, continuity of the contaminant source, and the physical and chemical characteristics of the contaminants. Low-density leachate entering the ground may float on top of the unconfined aquifer, whereas chemicals of a higher density sink. Contaminants in groundwater are not subject to dilution by the entire body of water as they are in surface water. Rather, dilution occurs very slowly by dispersion, which is produced by geologic variations in the path of the flow.

THE PROCESS OF RISK ASSESSMENT

Although some hazardous waste sites can pose a significant threat, the mere presence of a waste site does not mean that human health is imminently endangered. In some cases, the waste has not yet migrated into the environment. In others, the waste may never present a danger because of the nature of the chemicals present, the environmental fate of the chemicals, their migration potential, or the vulnerability of the local population. It can rarely be determined unequivocally that a waste site presents a threat to human health. Instead attempts must be made to assess the degree to which a specific waste site threatens an identifiable population or the environment. This type of investigation is known as risk assessment. A thorough assessment of risk requires expertise in several areas of science and engineering including chemistry, geology, hydrology, toxicology, and epidemiology. Five general criteria have been developed to conduct a thorough risk assessment of a hazardous waste site.

Site Description

The first step in assessing the risk presented by a waste site is a preliminary characterization of that site. This step is done quickly and provides general information about the operation of the site and its associated hazards. Records about the operation and management of the site are sometimes available and may provide valuable information on the types and amounts of waste and its location. The chronologic history of the waste-related operation may be useful in determining how far waste has migrated from the site and the extent of exposure of nearby residents. The records may indicate if waste was indiscriminately dumped into surface water or onto the ground, mixed, burned, or buried. Drums and tanks are examined to determine their physical integrity. Existing hydrogeologic data are reviewed to determine the potential routes of contaminant migration. Particular attention is paid to current and potential uses of surface water and groundwater, including the identification of all wells in the area. Estimations of the population at risk are made. Based on information gleaned from this initial survey, a determination is made concerning the need for a more detailed study of the site, beginning with an extensive survey and analysis of the waste on the site.

The extent to which the chemical makeup and, hence, the hazardous nature of the waste can be characterized varies greatly from site to site. If one or only a few waste streams are present at the site, identifying and quantifying the chemicals present is relatively straightforward. If hundreds of waste streams are present in tens of thousands of unmarked drums or mixed together in larger tanks, complete identification of the individual chemicals becomes very time consuming and expensive. In these cases quick screening tests are performed to determine the general chemical and physical characteristics of the waste. The parameters examined include pH, flammability, reactivity with water, oxidation and reduction potentials, vapor concentrations, and radioactivity. Based on these tests, incompatible wastes are segregated for safe storage and similar wastes can be combined for more detailed analysis.

Determining if a particular waste product is hazardous depends on several factors. For purposes of regulatory control, a waste is considered hazardous if it possesses at least one of the following characteristics to a specified degree: corrosivity, ignitability, reactivity, or toxicity. A waste is defined as corrosive if it has a pH of less than 2 or more than 12.5, or if it corrodes steel at a rate greater than 0.25 in. per year. An example of a corrosive waste is the spent pickle liquor from steel finishing operations. A liquid waste is considered ignitable if it has a flash point of less than 60°C (140°F). The flash point is the lowest temperature at which a liquid will give off enough flammable vapor so that in an intimate mixture with air and a spark it will ignite. A nonliquid is defined as ignitable if it is capable, under standard temperature and pressure of causing fire through friction, absorption of moisture, or spontaneous chemical change. Many spent industrial solvents are hazardous because of their ignitability. These solvents include acetone, methanol, xylene, toluene, and methyethyl ketone (MEK).

A waste is considered reactive if it possesses any one of the following properties: (1) it is normally unstable and readily undergoes violent change without detonation; (2) it reacts violently with water or forms potentially explosive mixtures with water or generates toxic gases or fumes when mixed with water; (3) it is a cyanide- or sulfide-bearing waste, which when exposed to pH conditions between 2 and 12.5, generates toxic gases or fumes; (4) it is capable of an explosive reaction at standard temperature and pressure or when heated. An example of a reactive waste is the wastewater treatment sludge from the manufacturing of explosives. Specific reactive chemicals include nitroglycerin and picric acid.

Waste is considered acutely toxic if it contains constituents that have been found to be fatal to humans in low doses or, in the absence of human data, it has been shown in animal studies to have an oral LD50 for the rat of less than 50 mg/kg, an inhalation LC50 in the rat of less than 2 mg/liter, or a dermal LD50 in the rabbit of less than 200 mg/kg. Examples of acutely toxic waste streams are those containing cyanide, arsenic, strychnine, or tetraethyl lead. Wastes are also considered toxic if they have been shown to have carcinogenic, mutagenic, or teratogenic effects on humans or other life forms. The form of the waste and its manner of disposal are also important in assessing the hazard. Liquid and gaseous wastes are more susceptible to rapid release into the environment than solid waste. Although wastes contained in intact drums or tanks are less prone to release into the environment, they are usually more concentrated than wastes found open to the atmosphere in surface impoundments.

Complete characterization of the waste confined to a site may not be necessary if the site is fenced and secured to limit the extent of human contact with it. It is necessary, however, to make a complete qualitative and quantitative analysis of all waste constituents that have migrated from the site. An extensive survey of the groundwater, surface water, and air surrounding the site must be performed. Individual chemicals and their concentrations in the affected medium must be known in order to assess the potential impact on the nearby population. As long as the waste remains on the site, the environment should be examined regularly even if previous tests showed that no contaminants had escaped.

Identification of Potential Migration and Fate of Waste Constituents

The risk assessment must determine the potential for migration of waste constituents from the site and predict the environmental fate of the escaping chemicals. The migration and fate of the chemicals released into the air, surface water, and ground water are determined by the physical and chemical properties of the waste and of the environmental medium into which they have escaped. Depending on these properties, certain physical, chemical, and biologic processes will predominate in controlling the migration and fate of the chemicals. These processes are complex and often interrelated. Important intrinsic properties of the migrating chemicals include their solubility in water, octanol-water (o/w) partition coefficient, vapor pressure, and specific gravity. The water solubility of a chemical determines the extent to which it dissolves in water; therefore, a highly soluble chemical is more likely to distribute in surface water, and contaminate and move with ground water. The o/w defines the extent to which a chemical will partition itself between an organic phase and a water phase. In surface water, the fatty tissue in fish may represent the organic phase. Consequently, the o/w can be used to predict the extent to which a chemical will bioaccumulate. In ground water, the o/w can be used to predict the extent to which chemicals will absorb onto organic soil particles to abate their migration. Vapor pressures define the extent to which a chemical in a liquid or solid phase will become a gas. The specific gravity of a chemical is the ratio of its weight to the weight of an equal volume of pure water. If a chemical is more dense than water, and insoluble in water, it will sink and form a separate phase that may move in a direction and at a rate different from that of the ground water. Conversely, if it is less dense than water, it will float on top of the unconfined aquifer.

Environmental factors affecting the migration of waste constituents include the soil type, geologic and hydrologic parameters, and the climate and topography. Finely textured soil such as clay and glacial till allow much less water movement than coarser soils such as sand and gravel, which provide rapid movement of water and other contaminated liquids. The location, size and depth of aquifers as well as the condition of aquitards between aquifers affect the extent to which ground water will be contaminated. The depth of the unsaturated zone lying between the land surface and the unconfined aquifer affects the rate at which contaminants enter the ground water. A deeper unsaturated zone reduces the rate at which the ground water is contaminated. Contaminants that are soluble in water move more rapidly in the saturated zone of the aquifer whereas insoluble contaminants, such as most pesticides, move slowly underground.

Climate is relevant because precipitation is the primary source of natural recharge to an unconfined aquifer. Soluble contaminants are likely to be carried with the precipitation if it first passes through exposed or buried waste before reaching the aquifer. Sites located in areas of high precipitation and high winds are more likely to release contaminants than those in arid or low wind areas. Temperature is also important because it affects the volatility of many organic solvents. Topography determines the likelihood of surface water drainage and flooding.

Naturally occurring physical, chemical, and biological processes may act to abate the migration of chemicals by attenuating their movement or promoting chemical changes to them. The physical and chemical processes include adsorption of contaminants to soil particles or suspended or bottom sediments in surface waters; photodegradation in air and surface water; chemical transformations, such as oxidation, reduction or hydrolysis; cation exchange of toxic metals with nontoxic cations on soil particles; and extraction of organic chemicals by organic soils, such as peat. Biologic processes include microbial degradation of organic compounds, microbial transformation of compounds containing sulfur or nitrogen, and microbial fixation of metals. These biologic processes are more common in the oxygen containing environment of the atmosphere or surface water than in the oxygen depleted anaerobic environment underground.

Evaluation of Exposure Potential

The next step in the risk assessment process is the exposure evaluation. Here estimates are made concerning the current and future population exposed to contaminants from the site, and the routes of exposure and its duration. As usual the elderly, the young, pregnant women, and the chronically ill are presumed to be more susceptible to the adverse health effects resulting from exposure to hazardous chemicals. A quantitative determination of the duration and intensity of exposure is far more elusive. Because these assessments are usually begun long after exposure has occurred, it is rarely known when individuals were initially exposed, what the concentration of contaminants was, and whether the exposure was continuous or intermittent. When this information is lacking, the duration and maximum levels of past exposures must be estimated based on chemical migration data and proximity of individual residents to the hazardous waste site. Measurable amounts of contaminants in body fluids or tissue may also indicate exposure. Body fluids such as blood, urine, milk, and skin oil or fat tissue can be analyzed, but the latter is rarely sampled because of the necessity for intrusive access.

Route and dose are the most important factors when evaluating the potential toxic effects of exposure to contaminants. Relatively low level exposures may produce more insidious chronic or long latency illnesses that may take months or years to produce symptoms. Exposure to high doses generally produces more prompt symptoms, such as skin irritation, nausea, vomiting, depression of the CNS, organ damage, or death. Ingestion of chemical contaminants may produce damage to the internal organs of the body, especially the liver and kidneys where most metabolism and elimination occur. Entry into the body by any route may produce acute or chronic toxicity at any number of sites in the body, depending on the chemical involved, its dose, and the health of the exposed individual.

Toxicological Evaluation

If it has been determined that people have been exposed or are likely to be exposed to contaminants escaping from a hazardous waste site, the next step in the risk assessment is a toxicologic evaluation. This involves a thorough review of the scientific literature pertaining to the chemicals found and their anticipated degradation products. Important information needed for this evaluation includes the physical and chemical properties of the agents, pharmacokinetic data on absorption, distribution, metabolism, and excretion, and known adverse health effects including acute, subacute, and chronic organ toxicity, teratogenicity, mutagenicity, and carcinogenicity. Data on the health consequences from human exposures are most useful in assessing the hazard from chemical contaminants, but human data are rarely available. If human exposures have been reported, they generally involve an acute exposure or a large dose of a single chemical, whereas most exposures from hazardous waste sites involve low doses of a mixture of chemicals. Because data on human exposure is so limited, considerable emphasis is placed on animal data obtained under controlled laboratory conditions.

Human Impact Assessment

Perhaps the most difficult aspect of assessing human risk from a hazardous waste site is determining what concentration constitutes an acceptable exposure to a particular chemical, or in a popular phrase, "How safe is safe?" It is generally accepted that substantial efforts should be made to eliminate, or at least to limit, human exposure to chemicals migrating from a hazardous waste site. If exposure has already occurred or if future exposures are unavoidable, one needs to be apprised of the ramifications and science is expected to supply the answers. Unfortunately, accurate determinations of the nature and extent of human illness based on known concentrations of contaminants in the environment are virtually impossible to make. Instead, mathematical models have been developed to predict the likelihood of illness. These models are based on extrapolations from animal studies and assumptions about the exposed individual. Most models assume that the exposed person is a healthy 70 kg male who consumes two liters of contaminated water per day. Other models make additional assumptions about ingestion of contaminated fish and some allow for variation in weight. No model, however, can account for the tremendous human variability in the exposed population. Many exposed individuals are reluctant to trust these models to predict the danger associated with their exposure.

Models to Predict Human Impact

Two kinds of models are used for estimating health risks caused by exposure to toxic chemicals: one for noncarcinogens and one for carcinogens. See Chapter 8 for the noncarcinogen model. Most models currently used to assess the endangerment from exposure to suspected carcinogens are based on the theory that thresholds do not exist for carcinogens; therefore, no safe level can be determined. Instead these models produce carcinogenic potency factors or unit cancer risk (UCR) values, which are used to estimate concentrations that correspond to statistical lifetime cancer risks. A UCR provides a calculated estimate of the additional incidence of cancer expected in an exposed population. For example, one model calculates a UCR of 10^{-6} for benzene at a concentration of 0.67 μgm/l. A UCR of 10^{-6} indicates a probability of one additional case of cancer for every one million people exposed. Because the nonthreshold model is linear to infinitely low concentrations, the additional lifetime risk is directly proportional to the concentration. Therefore, a benzene concentration of 6.7 μgm/l indicates one additional case of cancer per 100,000 people exposed. Calculated cancer risk levels are relatively abstract values and their correlation to the actual degree of risk to human populations exposed to low concentrations of environmental carcinogens is unknown. Even so, they will probably be used to predict the risks of exposure to carcinogenic contaminants from hazardous waste sites for some time to come.

The final step in risk assessment is the actual human impact evaluation. Sometimes attempts are made to identify actual injury or illness resulting from exposure to contaminants from a hazardous waste site. Several avenues exist to determine if an increased prevalence of illness exists because of exposure to contaminants. These avenues include epidemiologic studies, clinical examinations, interviews with local medical personnel, registries of potentially exposed persons, and reviews of vital statistics. (See also Chapter 20 for ways of assessing the risks of carcinogens, mutagens, and teratogens.) This risk assessment procedure focuses on the potential impact of hazardous waste on human health. Additional assessments are often performed to determine deleterious effects on aquatic organisms and wildlife. In many cases their health endangerment is greater than that of humans.

REMEDIAL ACTION—"HOW CLEAN IS CLEAN?"

When it has been determined that a hazardous waste site presents an imminent danger to human health, attempts may be made to eliminate or at least abate the hazard. Ideally, the waste should be removed for safe and ultimate disposal, and the surrounding environment cleansed to its original condition. However, because of the high costs and technical feasibility associated with such cleanup operations, this is often not possible or practicable. For some sites, the only feasible option is better containment of the waste on the site, and regular monitoring of the air, surface water, and groundwater to establish whether contaminants are escaping from the site. When remedial actions are undertaken to eliminate the hazard, alternatives specific for preventing contaminant migration from the site and for decontamination of soil and groundwater are considered.

Some waste and contaminated soil can be encapsulated on site by surrounding it with impermeable walls of clay or sheet metal and covering it with an another layer of clay. Alternatively, the waste and soil can be removed for more secure disposal elsewhere. Groundwater can sometimes be cleansed through the placement of purge wells into the contaminated aquifer. Water pumped from the aquifer is treated to remove contaminants and then reinjected into the ground. For moderately to severely contaminated ground water this process could take years to achieve cleanup of the aquifer to safe levels, and in some cases this level of decontamination cannot be attained. The costs associated with this degree of cleanup can be enormous. Some experts have argued that once an aquifer is contaminated beyond safe levels, the best option is to consider it lost for potable use rather than spend the millions of dollars necessary to clean it.

Ultimately, attempts are made to dispose of the waste from sites undergoing remedial action. The waste can be treated or encapsulated on site, transported to a more secure disposal site, destroyed by incineration or any of several other chemical, physical, or biologic processes, depending on the specific waste. Stabilization and disposal on the site may be practical for solid waste contaminants, which are relatively insoluble. However, it is rarely a practical solution for liquid or semisolid wastes. Waste from many sites undergoing remedial action have been taken to secured landfills. These are vaults surrounded by highly impermeable material to prevent the escape of wastes contained within the landfill. Some, however, that were built prior to 1981 are already leaking their contents. Obviously, these "secured" landfills will not serve as ultimate waste disposal sites. However, they are serving to consolidate waste from many sites into one, albeit they do not provide a final solution. As of 1986 land disposal of many types of waste, particularly liquid wastes, is forbidden by law.

Incineration is a possible method of ultimate disposal. New incineration technologies are being developed to provide for the high temperatures and gas scrubbing chambers necessary to completely detoxify many types of hazardous waste. However, many communities are reluctant to grant the required permits; the mentality is one of NIMBY (Not in My Back Yard). Injection of liquid wastes into deep wells or brine aquifers already containing water unsuitable as a potable source has been used to dispose of liquid waste.

The creation of new hazardous waste sites will be greatly reduced through an act passed by Congress in 1976. This is the Resource Conservation and Recovery Act (RCRA). RCRA provides for the tracking of hazardous wastes from the point of production to its ultimate safe disposal. This so-called cradle to grave tracking is controlled by regulations pertaining to the treatment, storage, transportation, and disposal of hazardous wastes produced in the U.S. Public and industry awareness of the consequences of haphazard waste disposal should also help to alleviate the creation of future problems. Although steps are now being taken to prevent the creation of future waste sites, it is clear that the dangers of past hazardous waste disposal and the uncertainties surrounding the assessment of this endangerment will be with us for some time into the future. These problems will continue to challenge many fields of science, including environmental toxicology.

SUMMARY POINTS

The United States produces more than one ton of hazardous waste for every U.S. citizen each year.

In 1980 Congress passed a law (CERCLA) that provides for a "superfund" to cleanup abandoned hazardous waste sites.

There are over 1100 waste sites on the National Priority List that may require some remedial action to abate endangerment to human health or the environment.

Contaminants may enter the groundwater from hazardous waste sites and be present without detection for years at concentrations exceeding standards set to protect human health.

It may not be possible to decontaminate moderately to severely polluted groundwater in order to make it a safe drinking water source.

The hazardous waste site risk assessment process includes the following steps: site description, identification of potential migration and fate of waste constituents, evaluation of exposure potential, toxicological evaluation, and human and environmental impact assessment.

Precise determinations of the nature and extent of human illness based on known concentrations of contaminants in the environment are virtually impossible to make.

Hazardous waste from uncontrolled sites can be stored on site and with appropriate environmental monitoring moved to secure landfills or possibly destroyed.

Congress enacted a law (RCRA) to provide for "cradle to grave" tracking of hazardous wastes, thereby greatly reducing the creation of new uncontrolled waste sites.

BIBLIOGRAPHY

Andelman JB, Underhill DW: Health Effects from Hazardous Waste Sites. Chelsea, MI, Lewis Publishers, 1987.

Crawford M: Hazardous waste: where to put It? Science 235:156, 1987.

Ember LR: Uncertain science pushes Love Canal solutions to political, legal arenas. Chem Eng News 58:22, 1980.

Heritage J: The new superfund: protecting people and the environment. EPA J 13:1, 1987.

Maugh TH: Just how hazardous are dumps? Science 215:490, 1982.

Paustenbach DJ: The Risk Assessment of Environmental and Human Health Hazards, A Textbook of Case Studies. New York, John Wiley and Sons, 1989.

Phillipps AM, Silbergeld EK: Health effects studies of exposure from hazardous waste sites: where are we today? Am J. Ind Med 8:1, 1985.

U.S. Environmental Protection Agency: Risk Assessment Guidance for Superfund: Volume 1, Human Health Evaluation Manual (Part A). EPA/540/1-89/002. Washington, DC, U.S. EPA, 1989.

19

CHAPTER

THE ALCOHOLS

Organic chemicals that have a hydroxyl group (-OH) attached to a carbon atom in a straight or branched chain of carbons are called alcohols. The simpler alcohols (Fig. 19–1) have similar physical and chemical properties in that they are all clear, colorless liquids that will ignite and burn if exposed to an open flame. Most have only faint odors and no real taste; a mild burning sensation may be noted if placed in the mouth, on the tongue, or in the stomach if swallowed. They are freely miscible in all proportions with water. If consumed in appropriate doses, all are capable of producing central nervous system (CNS) depression and the familiar syndrome of inebriation. Several have important additional toxic effects that are unrelated to the CNS depression. Hundreds of other alcohols are known, but the four illustrated in Figure 19–1 are the most important commercially and toxicologically.

The simplest alcohol that can exist is methanol or methyl alcohol (Fig. 19–1). Methanol can be prepared by heating wood to high temperatures without burning it and separating the alcohol from the other volatile products, hence its other synonym, wood alcohol. Methanol is less potent as a CNS depressant than longer chain alcohols, but if enough is consumed, it will produce inebriation. Indeed, it may have been the origin of the expression "blind drunk." Unlike any other alcohol, methanol poisoning can result in damage to the retina of the eye. If the victim is treated in time, the damage may be fully reversible. In many unfortunate cases, however, permanent total or partial blindness has resulted. The eye effects occur late in the poisoning usually after recovery from the initial inebriation. The first signs may be complaints of blurring or dimness of vision, accompanied by an increase in the acidity of the blood (metabolic acidosis). These manifestations of methanol toxicity are unique to primates, i.e., man and some but not all species of monkeys.

The inebriation is probably caused by methanol itself, but the retinal damage and metabolic acidosis are caused by a biotransformation product, formic acid. The administration of baking soda (sodium bicarbonate) can reverse the acidosis, but it will not prevent blindness. Measures are needed to prevent the biotransformation of methanol to formic acid. The first effective chemical

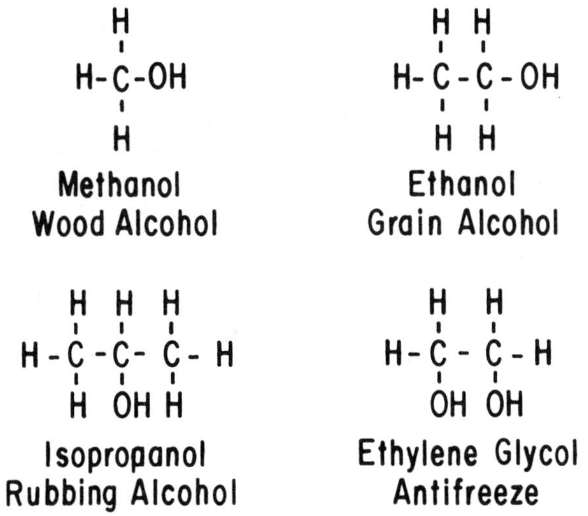

Figure 19–1. Four simple alcohols of commercial and toxicologic importance.

discovered that could do this was ethanol or beverage alcohol. Although it seems paradoxical to give one alcohol to prevent the toxic effects of another, the rationale is scientifically sound and the treatment has been proven to be effective. Methanol and ethanol are biotransformed by the same enzyme, but ethanol is not converted to more toxic forms. By continuously or repeatedly giving ethanol, one can competitively slow or prevent methanol conversion to formic acid, and the methanol will be excreted unchanged in the urine.

Methanol has been marketed as an emergency substitute for gasoline (Rescue), as a fuel for cooking outdoors (Sterno), in windshield washer de-icer fluids, and in cold weather gasoline additives to prevent freezeup. During World War II, methanol was one of the ingredients used in torpedo fuel. Epidemics of poisonings occurred among military personnel who had access to "torpedo juice."

Ethylene glycol (Fig. 19–1) could be regarded as a double alcohol because it has two hydroxyl groups, one on each carbon. It is the major ingredient in automobile antifreeze. Like methanol it is biotransformed into acidic and more toxic forms. Although its ingestion by humans has not been associated with blindness, the same measures used in methanol poisoning are effective against ethylene glycol poisoning and for the same reasons, namely, to prevent the formation of more toxic biotransformation products. Some of the biotransformation products of ethylene glycol are directly toxic to the kidney, and late acute renal failure is a common consequence of the ingestion of toxic amounts.

Isopropanol (Fig. 19–1) or rubbing alcohol is more potent as a CNS depressant than ethanol. If large amounts are consumed, the coma produced can be so profound that the blood pressure falls and blood flow to the kidneys may be inadequate, resulting in acute renal failure as a late and indirect consequence. Isopropanol also undergoes biotransformation by the same pathway as the other alcohols, but the end product, acetone, is no more toxic than isopropanol. Therefore, there is no reason to attempt to interfere with its biotransformation in poisoned patients. Such patients, however, may be found in a coma with an acetone odor on their breath. It would be easy to confuse such patients with diabetics who have not taken an adequate dose of insulin. Unlike diabetics the isopropanol-poisoned patient will not be passing large amounts of sugar in his urine.

Ethanol has many important commercial uses in addition to its ubiquitous distribution in the form of alcoholic beverages. It is an important solvent and vehicle in numerous types of products some of which are, and some of which are not, intended for human internal use. Many liquid prescription and nonprescription drugs contain ethanol as an "inert" ingredient. If a medicinal product is called an elixir, spirit, or tincture it almost invariably contains ethanol. Such products should be avoided by former alcoholics or individuals receiving a drug used in the aversion therapy of alcoholism, Antabuse (see below). Some mouthwashes contain surprisingly high concentrations of ethanol, as do aftershave lotions, perfumes, spray deodorizers, shoe and floor polishes, and some liquid waxes.

Beverage alcohol is heavily taxed. At least one-third of the price of a bottle of whiskey (more in some states and less in others) consists of taxes which constitute an important source of revenue for local, state, and federal governments. Even to our government, however, it did not seem fair to tax equally ethanol that was going to be used in products that were never intended for internal consumption or recreational use. Therefore, regulations were passed that permitted the deliberate addition of adulterants to beverage ethanol that would make it unfit, or at least undesirable, for human consumption. The mixtures are called denatured alcohols, and there are many different recipes depending on the type of product in which they are to be used. The adulterants are often selected on the basis of a foul taste or offensive odor, but they are also usually present in such minute concentrations that they make no contribution to the intoxication syndrome if denatured alcohol is consumed instead of beverage ethanol. Shellac thinner may be an exception. Depending on the particular product, it may or may not be denatured with large amounts of methanol which could give rise to the problems described above if ingested in large amounts. Denatured alcohol, isopropanol, and ethylene glycol are all readily available at prices considerably below their inebriating equivalents in beverage ethanol, and for this reason they are often consumed in desperation.

Ethanol for commercial uses other than beverages can be made synthetically, but by law all alcoholic beverages must be produced by the natural process of fermentation. Many natural materials can serve as the starting points for the production of alcoholic beverages as long as they contain adequate amounts of sugar (glucose or sucrose). Familiar examples are the wine from grapes and hard cider from apples. Cereal grains from which we produce rye whiskey and corn liquor do not contain sugar as such, but they do contain a sugar polymer in the form of starch. Before fermentation can take place, the starch must be broken down into sugar. Ground cereal grain is mixed with water to make a mash. Malt is then added to break the starch down into sugar. Malt is also prepared from cereal grains by moistening it and allowing the seeds to sprout. During the sprouting process, the seeds make the enzyme that breaks starch down into sugar, beta-amylase. The sprouts are dried at a low temperature, which stabilizes the enzyme, and ground into malt that contains many other things, but the beta-amylase is the key player. Sometimes the malt is roasted at a higher temperature, which gives it a darker color and results ultimately in a darker, sweeter beer called bock beer. Certain vegetable pulps are also used to make alcoholic beverages, such as potatoes, which ultimately yield vodka, and cactus pulp, which is used to make tequila. The starting material for rum is molasses.

The actual fermentation is carried out by yeasts, which break down the sugar into ethanol and carbon dioxide (CO_2). If the fermentation is carried out in a sealed container, the CO_2 will remain dissolved in the beverage resulting, in champagne and other sparkling wines. Fermentation for beer is not done in sealed containers and most of the carbon dioxide is lost to the atmosphere. Additional CO_2 is added artificially from compressed tanks just prior to sealing the beer in its market container.

The mash is passed through a filter to remove the ground cereal grain, or alternatively, the liquid on top, called the wort, can be siphoned off leaving the grain behind. The wort can be used to make either beer and its related beverages or it can be used to make whiskey and its cousins. In making beer the next step is to add hops. Hops come from the cones of a common vine, and it gives beer its slightly bitter taste. The beer is then lagered, which simply means that it is stored in a cool environment to mature and clarify. Finally, most beer intended for bottles or cans is pasteurized by heating it to destroy the yeast cells and prevent further fermentation. Draft beer in barrels and kegs is not pasteurized, which may account for some subtle differences in taste or other qualities. So-called draft beer in bottles and cans has been subjected to a cold pasteurization process, in which it is passed through very fine filters to remove the yeast.

When the concentration of ethanol reaches about 16% by volume during fermentation, the yeast cells are killed. Like humans, yeast can only tolerate so much ethanol before they fall victim to their own folly. For that reason, 16% by volume ethanol is the strongest alcoholic beverage one can prepare by fermentation. It is not by accident that this is about the average ethanol content of wines. In order to make hard liquors such as gin, rum, brandy, vodka, bourbon, or scotch, the products of fermentation must be distilled, hence the name, distilled spirits. Stills may differ widely in physical appearance, but all need the same basic elements: (1) a source of heat, (2) the still proper or boiler, (3) a condensor to condense ethanol vapors back to the liquid form, and (4) a container to receive the liquid ethanol. As the fermentation liquid is heated, ethanol begins to boil off at 78°C, and if the temperature is kept below 100°C, the boiling point of water, the ethanol can be partially or almost completely separated from the water depending on the efficiency of the still. A good chemical laboratory still can make 95% ethanol by volume. Absolute ethanol is 100% by volume, but it will absorb water vapor from air to go to 95% if it is not stored in a sealed container. An alternative way of concentrating the products of fermentation is by freezing. Homemade applejack may be placed outside in freezing temperatures, and when any ice formed is removed, the remaining liquid is more concentrated in ethanol. The temperatures, however, must be low indeed because ethanol can also function as an antifreeze.

Figure 19–2 shows the ethanol content of common beverages and illustrates the proof system for expressing alcohol content. Ordinary American beers are about 6% ethanol by volume. Some states authorize the sale of beer containing about half that concentration to minors, and today "beers" almost totally devoid of alcohol are available. Closely related beverages are ales, porter, stout, and malt liquor. It is obvious that these represent beverages in which fermentation has been artificially stopped before it went anywhere near completion. Other countries make beers of higher ethanol content. Indeed, it should be possible to make a beer that is 16% ethanol by volume.

By coincidence, grapes contain about 16% sugar by weight, which corresponds to the ethanol content by volume of wine. If fermentation goes to completion, most or all of the sugar is consumed, resulting in a "dry" wine. If fermentation is stopped short of completion, such as at 10 to 12% ethanol, the product is a sweet

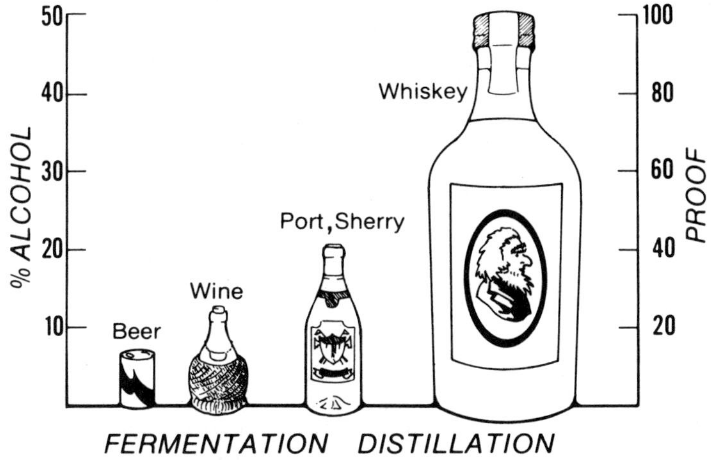

Figure 19–2. The alcohol content of some common beverages as a percentage by volume and its equivalents in the proof system. (From Smith RP, Barrington JM: Unit 1. The biochemistry and pharmacology of alcohols. *In* The Joan B. Kroc Foundation: Alcohol Use and its Medical Consequences. Timmonium, MD, Milner-Fenwick, 1981.)

wine or dessert wine. When wine is distilled, the product is dry brandy to which sweet fruit syrups (e.g., apricot, peach, ginger) may or may not be added. Most distilled spirits are about 40 to 50% ethanol, which indicates that the stills used are not very efficient. So-called fortified wines, such a sherry or port, are made by mixing together wine and brandy to achieve an ethanol content of about 20 to 25%.

The proof system was derived simply by doubling the percentage of ethanol by volume in beverages such that absolute ethanol is 200 proof and most distilled spirits are 80 to 100 proof. The system is said to have originated from a former criterion of the potency of distilled spirits. The product of a given still was placed in a shallow basin and mixed with gunpowder. In response to the tossing in of a lighted match, there was sometimes an audible "proof." It happens that the minimum ignitable concentration of ethanol in water is about 50% by volume. Because it was thought at the time that distilled spirits represented the maximum ethanol concentration that could be achieved, it was assigned the value of 100 proof.

Ethanol readily passes through all membranes of the body by diffusion (see Chapter 6) with two exceptions: the intact human skin and the urinary bladder. When inhaled, ethanol vapors in air are rapidly absorbed systemically and with a high efficiency. The inhalation of 0.007% ethanol vapor in air by volume could result in a blood alcohol concentration (BAC) of 0.045% within 3 to 4 hours in an adult human. That concentration of ethanol vapor, however, is extremely irritating to the eyes and the odor is obvious. It is almost inconceivable that such an exposure could go unnoticed by a normal human. Concentrations of 0.009% by volume are intolereable to most humans, and they would not voluntarily remain exposed. A forced exposure to 0.9% for an hour could result in death. In a number of tragic cases infants have been given over zealous alcohol (ethanol or isopropanol) rubs to reduce fever, and left in a confined area to become seriously or fatally poisoned.

The most rapid way of raising the BAC is to inject it intravenously. Dilute solutions are used in that way to treat victims of methanol or ethylene glycol poisoning. If inserted into the bowel as an enema, ethanol will also be systemically absorbed, but one is at a loss to explain why such a route might be selected.

By definition, alcoholic beverages are taken by mouth, and a small fraction of the ethanol is absorbed through the mucous membranes lining the oral cavity. After the bulk of it is swallowed, any traces of ethanol remaining in the mouth are completely absorbed within 20 minutes. For this reason police officers are instructed to wait at least 20 minutes after apprehension before administering an alcohol breath test. As with most chemicals, absorption of ethanol from the stomach is less efficient than from the first 8 to 10 inches of the small bowel, the area called the duodenum. On average, only 25% of an ingested dose is absorbed from the stomach, whereas 75% is absorbed from the duodenum. Therefore, it can be seen that the rate of ethanol absorption is strongly influenced by the gastric emptying time.

Gastric emptying is controlled by a valve called the pylorus. When food is consumed, the pylorus closes and remains closed until digestion is complete. If ethanol is consumed along with or shortly after meals, its rate of absorption is slower than when it is consumed on an empty stomach. Thus, the admonition against drinking on an empty stomach has a solid scientific basis, and the rapidly disastrous consequences are well known. Because it takes longer to digest fats than proteins or carbohydrates, this effect may be more pronounced after a fatty meal. This phenomenon is also a part of the folklore, but the true explanation is not generally appreciated.

On average and after a single drink ethanol begins to enter the blood within 1 to 2 minutes and absorption is 90% complete within an hour. The pylorus is often removed during surgery for a peptic ulcer or cancer of the stomach. As a result, such patients acquire what is called the dumping syndrome because all consumed food and beverages pass rapidly into the small intestine. These people are intolerant to ethanol because it is so rapidly absorbed. Large quantities of concentrated alcoholic beverages are irritating to the stomach lining and may cause a spasm of the pylorus to slow absorption even on an empty stomach, or trigger vomiting, a common sign of overindulgence, or cause or aggravate a peptic ulcer.

All absorbed ethanol goes directly to the liver, but the blood eventually distributes it to the rest of the body. Ethanol distributes homogeneously in total body water. Water makes up 50 to 70% of total body weight, but it is distributed unevenly in various tissues and organs. Fat contains very little water, and consequently it will have low concentrations of ethanol. In general, women have a higher percentage of body fat per unit of body weight than men. In a woman and a man of equal weight, the same dose of ethanol in the woman will result in a higher BAC. The stomach, itself, has some capacity for metabolizing ethanol, and this activity is higher in men than in women, which also contributes to the higher BAC in women.

The body rids itself of ethanol almost exclusively by biotransformation. About 90 to 98% of the total ingested is converted to CO_2 and water. The biotransformation occurs in the liver, but cytochrome P450 is not importantly involved. Ethanol is an exceptional chemical in that the enzymes involved in its biotransformation are in the cell cytosol and mitochondria instead of the endoplasmic reticulum. The biotransformation of 1 ounce of 100 proof whiskey requires 1.0 to 1.5 hours, and results in the production of 100 calories. Ethanol and sugar are examples of calorigenic substances that have little or no nutritive value. Alcoholics derive a large share of their caloric intake from ethanol, and as a result may fail to consume an adequate diet so that nutritional disorders are often superimposed on the alcoholism.

Because ethanol is distributed in total body water, it is excreted in urine, tears, saliva, perspiration, and with water vapor on the breath. All of these excretory routes together, however, account for only 2 to 8% of the ingested dose. The level in the breath can be measured by the Breathalyzer or Intoximeter, instruments used in testing drivers for evidence of operating a motor vehicle under the influence of alcohol. Its concentration in the expired breath is proportional to its concentration in arterial blood.

Because alcohol passes through the liver on its way to the rest of the body, biotransformation begins immediately after absorption. When the rate of absorption exceeds the rate of biotransformation, the BAC will rise. When drinking stops, the rate of absorption will fall and eventually reach the rate of biotransformation so that the BAC reaches a plateau. Thereafter, the BAC will fall with time until ethanol is no longer detectable. Ethanol biotransformation is also unusual in that the BAC declines linearly with time. The plasma concentration of most drugs falls off logarithmically with time. The rate limiting enzyme for ethanol biotransformation is saturated over a wide range of plasma concentrations so that it is working as fast as it can, and there is no way to speed it up. Because it can only act on the same amount of ethanol per unit of time, its level of activity is a critical determinant of how much ethanol can be consumed per unit of time before reaching a critical BAC. The average 70 kg adult male can consume 1 ounce of 100 proof whiskey every hour for long periods of time without ever reaching a BAC of 0.1%. If the rate of consumption is doubled to 2 ounces per hour, however, the same subject would reach 0.1% in 2 to 3 hours. During the falling phase, the BAC declines at a steady rate of 0.015% per hour.

Many unsuccessful attempts have been made to find a drug or procedure that would accelerate the rate of ethanol biotransformation, Even the artificial kidney removes ethanol at a rate roughly equivalent to biotransformation so that the combination only doubles the rate of removal. This risky procedure must be reserved for those patients who are judged to be close to death.

Because the enzymes involved in ethanol biotransformation are not microsomal, they are not induced to the same degree as P450 isozymes would be when they are exposed to their inducers. The modest degree of induction, however, does account for the small component of metabolic tolerance (below). Oddly, the chronic intake of ethanol does induce P450 even though ethanol is not an important substrate for that system. For that reason chronic drinkers may biotransform some drugs at rates faster than abstainers. A further odditity is that ethanol can inhibit the biotransformation of some drugs by P450 although it is not itself a substrate. Perhaps it merely "sits" on the enzyme and prevents it from acting on other drugs. The significance of this phenomenon is that if ethanol and such drugs as sedatives are taken together, the biotransformation of the sedative to inactive products may be dangerously slowed. Thus, some drugs, especially CNS depressants, should not be consumed with ethanol. These possibilities for both the induction and the inhibition of cytochrome P450 conspire in a host of potential ethanol-drug interactions that sometimes result in unpredictable effects.

Conversely, several drugs are known to slow the rate of ethanol biotransformation. More or less specific inhibitors are known for both of the critical enzymes involved, alcohol and aldehyde dehydrogenases. The inhibitors of alcohol dehydrogenase are of interest for their potential in the management of methanol and ethylene glycol poisonings. The inhibitors of aldehyde dehydrogenase such as disulfiram have been exploited in the aversion therapy of alcoholism. This drug was discovered by accident when a Danish chemist who had synthesized it as a possible agent for use against intestinal worms went to a cocktail party and became violently ill after his first drink. He experienced widespread vasodilation resulting in a warm flush, a precipitous fall in blood pressure and a rapid increase in heart rate (palpitations). The experience suggested to him a new class of therapeutic drugs for use in alcoholism in which the disulfiram pretreated patient would avoid ethanol for fear of precipitating the reaction. The treatment is not without its own risks, however, since there have been reports of it resulting in collapse, shock, and even death in patients who were either unwittingly exposed to ethanol as the vehicle in a medicinal product or who were unable to maintain abstinence.

The ethanol-disulfiram reaction is presumed to result from the block in ethanol biotransformation and the accumulation of the intermediate, acetaldehyde, following an ethanol challenge. This hypothesis is somewhat oversimplified, but it is a reasonably good approximation. Chemicals related to disulfiram may be encountered as pesticides, industrial intermediates, and even as therapeutic drugs used for other purposes. Any suddenly developing intolerance to ethanol should provoke a search in the patient's environment for possbible exposure to such chemicals or drugs.

The use of alcoholic beverages dwarfs that for all other recreational drugs. If one defines an alcoholic as a person who has difficulty in keeping steady employment because of a drinking problem, the number of alcoholics in the U.S. is in the tens of millions. Cirrhosis of the liver, usually, but not invariably a result of alcoholism, is among the ten leading causes of death in the U.S. The annual consumption of distilled spirits in this country is well in excess of a half billion gallons. Statistics from other countries, however, suggest that we are not the most intemperate nation in the world.

Few people reading this page have not experienced first hand at least some of the effects of ethanol. Self-experimentation, however, is not among the more enlightening ways to learn about the effects of ethanol on the body. Its subjective effects are often manifested as a series of subtle paradoxes, and these are contributing factors both to its popularity and to the misery for which it is responsible. For example, low to moderate doses of ethanol produce an apparent and paradoxical stimulation of the CNS which is the so-called excitement or euphoric stage of inebriation. This is the "high" that is sought by the social drinker. Since ethanol is known to have only depressant effects on the CNS, this "stimulation" must be rationalized as actually caused by a depression of the highest inhibitory control mechanisms in the brain. This release of the brakes while the accelerator is still depressed results in a loss of inhibitions and self-restraint.

Many social drinkers believe that they are experiencing greater clarity of thought and deeper insights while under the influence, whereas in actuality there is a dulling of the finer grades of discrimination, memory, concentration, and judgement. The paradox of clear thinking disappears and is replaced by the hangover the following morning. Motor skills involving familiar and habitual tasks are less affected than work requiring higher reasoning. The involvement of ethanol in more than half of all automobile fatalities proves that appreciable decrements in such skills is the general rule given the consumption of an adequate dose.

The usefulness of ethanol as a general anesthetic is a myth perpetuated by western movies. Although it acts much like a general anesthetic such as might be used for major surgery, the anesthetic threshold lies too dangerously close to the threshold for the depression of breathing for it to be useful. Deeply intoxicated patients are often in need of artificial respiration. The modest increase in pain threshold may be ascribed to euphoria or an altered subjective response to pain.

A subjective feeling of warmth after drinking is due to increased blood flow to the skin and to the stomach. Ethanol also increases sweating which together with the vasodilation in the skin accelerates body heat loss. Thus, drinking when exposed to a cold environment is both irrational and dangerous, and it could result in severe hypothermia. Subjective impressions of an increased capacity for physical exertion are usually ascribed to a lessened appreciation of fatigue. The apparent sexual arousal results, of course, from the well known loss of inhibitions. The increase in urine formation is generally appreciated by the public, but it is not commonly recognized by lay persons as resulting from an effect on a gland in the brain that is then mediated through the kidneys. Ethanol blocks the release of a hormone from the pituitary gland that would ordinarily slow the rate of formation of the urine. Low concentrations of ethanol stimulate the secretion of digestive juices in the stomach; hence, the civilized custom of an aperitif before dinner. As already noted high concentrations irritate the stomach, and chronic gastritis is a common complaint among hard core drinkers.

As can be gleaned from above, most of the important effects of ethanol are mediated through the CNS. The analogy between the brain and an old fashioned telephone switch board is a gross oversimplification, but in some ways true. Messages are continually being received by the brain from other parts of the body, and instructions for the responses are continually being sent back over nerves. The familiar phenomenon of staggering when inebriated is not caused by an effect of ethanol on leg muscles or even on the nerves that control those muscles. It is the result of an effect of ethanol on the brain so that the messages sent out over the nerves to the leg muscles are garbled.

As the doses of ethanol are increased, the effects become less paradoxical and the inevitable progression from confusion, stupor, coma, and even to death clearly illustrate its depressant effects on the CNS. As shown in Figure 19–3, a fifth to a quart (a liter) of distilled spirits, if rapidly consumed, can be lethal to the average adult. Many deaths have resulted from foolish, macho drinking contests. Note that the amount of alcohol in an ounce of distilled spirits is equivalent to that in an average serving of wine or in a 12 ounce can of beer; its as if the servings had all been adjusted to provide the same dose of ethanol. Consumption of a six pack of beer over a short period of time is enough to carry one over into the state of legal intoxication.

Although by necessity the legal criterion for intoxication is a sharply defined value for the BAC, the stages of ethanol intoxication may occur at widely different BAC from one individual to another. In part this is a result of tolerance to the effects of ethanol. Individuals are born with different degrees of tolerance to ethanol. This is called inherent or innate tolerance, and the differences from person to person can be appreciable although the mechanism is unknown. The other kind of tolerance, acquired, can be superimposed on the inherent tolerance because it is acquired by drinking repeatedly and chronically.

There are actually two types of acquired tolerance, and in a practical sense it is difficult to distinguish the contribution of each

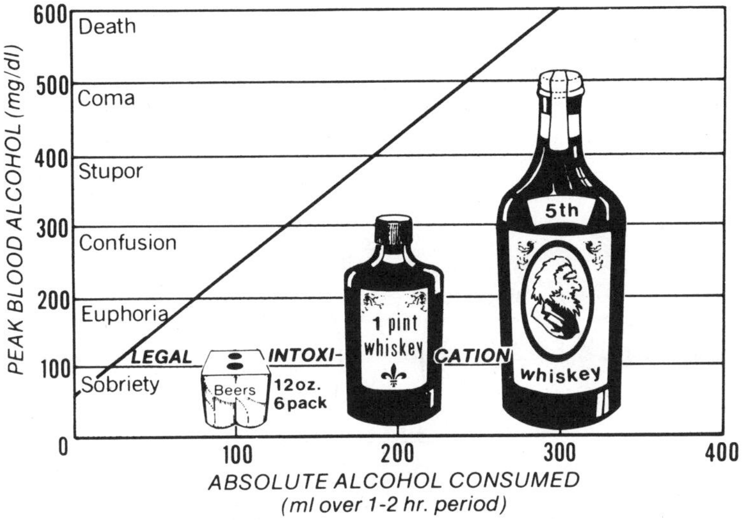

Figure 19–3. The relationships among the blood concentrations of alcohol, the amount of alcohol consumed (the dose), and the signs of intoxication (the effect). (From Smith RP, Barrington JM: Unit 1. The biochemistry and pharmacology of alcohols. *In* The Joan B. Kroc Foundation: Alcohol Use and its Medical Consequences. Timmonium, MD, Milner-Fenwick, 1981.)

to the total. These are illustrated in Figure 19–4. Metabolic tolerance was alluded to above. It is acquired as the body increases modestly its ability to biotransform ethanol. As a result the individual must consume more ethanol in order to reach the same BAC. The other type of acquired tolerance is called neuronal tolerance. It is acquired at the level of the nerves within the CNS, but nothing is known about the mechanism involved. Somehow the nerves learn how to function in a nearly normal fashion in the presence of concentrations of ethanol that previously would have produced signs of impairment. As a result of this kind of tolerance the individual must consume more ethanol in order to achieve the same effect. Some individuals can achieve surprisingly high levels of tolerance to ethanol; even so, on average the degree of tolerance is not nearly so great as that developed by the heroin addict to that drug. Unlike heroin, tolerance does not seem to develop to the lethal effects of ethanol. The danger of tolerance is that it leads to increased consumption and the risk of developing physical dependence.

Figure 19–5 illustrates the nerve membrane fluidity hypothesis about the above phenomena. It is postulated that in a single drinking bout ethanol interacts with nerve membranes in the brain to increase their fluidity, resulting in a garbling of the information they transmit and the appearance of the signs of inebriation. This process is rapidly reversible, but the recovery of the membranes may be associated with the symptoms of the hangover. With prolonged chronic consumption, however, the nerves acquire tolerance and remain fluid even in the absence of ethanol. After years,

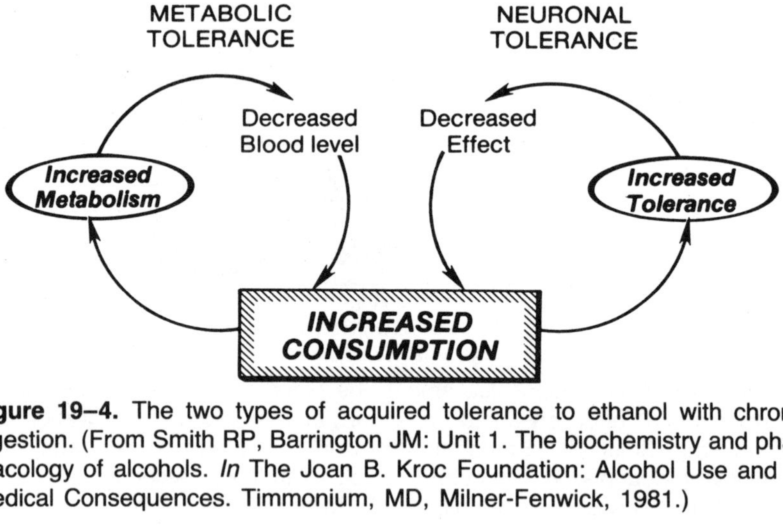

Figure 19–4. The two types of acquired tolerance to ethanol with chronic ingestion. (From Smith RP, Barrington JM: Unit 1. The biochemistry and pharmacology of alcohols. *In* The Joan B. Kroc Foundation: Alcohol Use and its Medical Consequences. Timmonium, MD, Milner-Fenwick, 1981.)

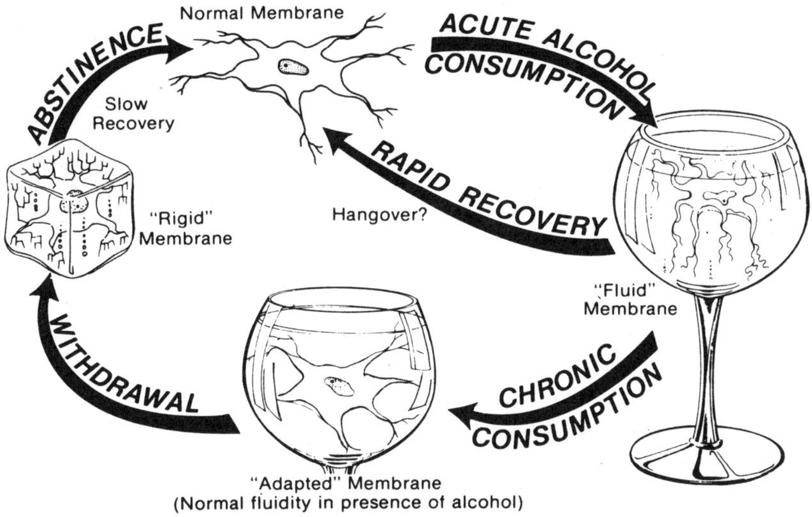

Figure 19–5. The neuronal membrane fluidity hypothesis about the effects of single drinking bouts, the hangover, and the development of physical dependence with chronic alcohol consumption. (From Smith RP, Barrington JM: Unit 1. The biochemistry and pharmacology of alcohols. *In* The Joan B. Kroc Foundation: Alcohol Use and its Medical Consequences. Timmonium, MD, Milner-Fenwick, 1981.)

physical dependence sets in so that when ethanol consumption is stopped the individual experiences a violent and dangerous abstinence syndrome in which the membranes become abnormally rigid. The abstinence syndrome is called delirium tremens after its two characteristics: delirium and tremors. The heroin abstinence syndrome is uncomfortable but it is not life-threatening. Ethanol withdrawal is potentially fatal.

The severity of the ethanol withdrawal reaction depends on the degree of physical dependence developed by the individual (Fig. 19–6). The danger period in a major withdrawal reaction is in the first two days when violent convulsions may threaten life. The specific drug for preventing or perhaps treating withdrawal is, of course, ethanol, but there is an understandable ethical reluctance to recommend or use it. Heroin and its relatives are of no value; there is no cross-tolerance between ethanol and heroin. However, there is cross-tolerance to some degree with a wide variety of CNS depressant and sedative drugs. The drug of choice today is diazepam (Valium), and the dose is tapered with time until the signs disappear.

Metabolic tolerance to ethanol in which P450 isozymes are induced accounts in part for cross-tolerance to CNS depressant drugs

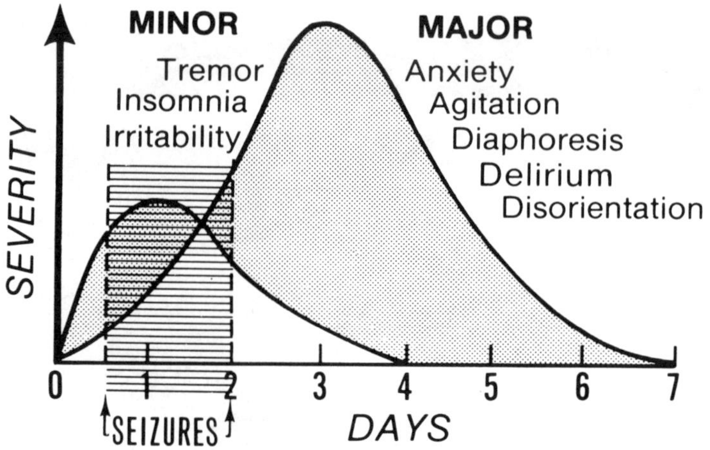

Figure 19–6. Patterns of signs and symptoms in alcohol withdrawal from a person who has become physically dependent. (From Smith RP, Barrington JM: Unit 1. The biochemistry and pharmacology of alcohols. *In* The Joan B. Kroc Foundation: Alcohol Use and its Medical Consequences. Timmonium, MD, Milner-Fenwick, 1981.)

that are inactivated by biotransformation. Neuronal tolerance to ethanol accounts for cross-tolerance to CNS depressant drugs that are not inactivated by biotransformation such as the gaseous general anesthetics, halothane and nitrous oxide.

An interesting and frightening hypothesis illustrated in Figure 19–7 suggests that tolerance to ethanol may begin with an individual's first drinking episode. When given tests to evaluate motor or intellectual function, people who do not habitually drink invariably perform better on the falling phase of the BAC curve than on the rising phase even though the BAC was the same in each case. This has been interpreted as evidence of the rapid development of tolerance (acute tolerance). As folklore would have it the proverbial "hair-of-the-dog" does provide relief from some of the symptoms of hangover. Thus, the hangover has been interpreted as a mild form of the abstinence syndrome. If these theories have validity both tolerance and physical dependence begin to build in an individual with the first drink, and if continually and chronically reinforced they will progress to long-lasting physical dependence.

In summary, although ethanol is not an environmental chemical in the sense that it is responsible directly for pollution, it is widely used, ubiquitously distributed, and can result in important interactions with drugs and other chemicals that are environmental pollutants. Lifestyle influences, such as smoking and

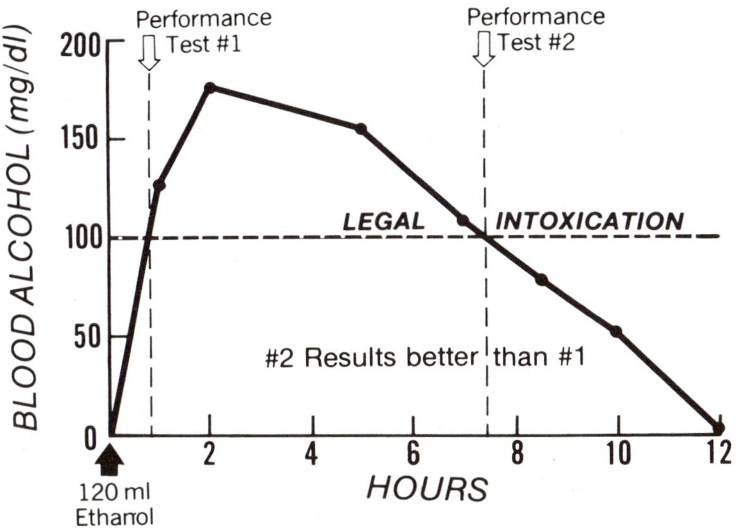

Figure 19–7. The evidence in support of the hypothesis for the development of acute tolerance to ethanol. (From Smith RP, Barrington JM: Unit 1. The biochemistry and pharmacology of alcohols. *In* The Joan B. Kroc Foundation: Alcohol Use and its Medical Consequences. Timmonium, MD, Milner-Fenwick, 1981.)

the use of ethanol and other recreational drugs, are responsible for more chemically induced disease than all environmental pollutants combined. Health and environmental problems seldom exist in isolation. It is the inter-relatedness of unsuspected things that must be perceived and appreciated before definitive corrective actions can be taken.

SUMMARY POINTS

Methanol and ethylene glycol are biotransformed to more toxic products resulting in inebriation followed by acidosis; the former also produces blindness and the latter kidney damage; ethanol antagonizes the effects of both other alcohols.

Ethanol is an important solvent for medications and other products not intended for internal consumption; denatured alcohol is used in the latter because it is not taxed; the denaturants or adulturants usually make no contribution to the effects of the ethanol.

Alcoholic beverages are made by fermentation of fruit or vegetable pulps or cereal grains (up to 16% ethanol), followed by distillation to hard liquors (40 to 50%).

SUMMARY POINTS (continued)

Ethanol passes through all biologic membranes except the intact skin and the urinary bladder; it is absorbed by all other routes including the lungs.

Absorption after drinking is strongly influenced by gastric emptying time; after absorption it distributes in total body water.

The effects of ethanol are terminated by biotransformation in the liver; P450 isozymes are not importantly involved, but ethanol both induces and inhibits P450.

Disulfiram inhibits ethanol biotransformation and results in an alarming reaction when an alcoholic subject challenges himself with ethanol.

The effects of ethanol are mediated almost exclusively through the CNS; it is a primary CNS depressant that can produce coma and death in respiratory failure.

Long term abuse of ethanol can result in metabolic and neuronal tolerance and physical dependence manifested by a dangerous abstinence syndrome.

BIBLIOGRAPHY

Smith RP, Barrington JM: Unit 1. The biochemistry and pharmacology of alcohols. *In* Alcohol Use and its Medical Consequences. Timmonium, MD, Milner-Fenwick, 1981.

Raul TW: Hypnotics and sedatives; ethanol. *In* Gilman AG, Rall TW, Nies AS, Taylor P (eds): Goodman and Gilman's the Pharmacological Basis of Therapeutics, 8th Ed. New York, Pergamon, 1990.

20

CHAPTER

CARCINOGENICITY, MUTAGENICITY, AND TERATOGENICITY

Carcinogenesis, mutagenesis, and teratogenesis can be regarded as special kinds of toxic effects that have some overlap (Fig. 20–1). A small group of chemicals are capable of eliciting at any one time each of the three types of responses. Other groups can elicit two of the three responses, and still other groups elicit only one of the three. A fourth response, immunosuppression, will not be considered here even though it could be represented as a fourth circle in Figure 20–1. Most chemicals, however, elicit none of these responses.

CARCINOGENESIS

Whether cancer is basically one disease with many faces or whether there are fundamental differences among the various forms of cancer is still unknown as are most of the causes of cancer. Chemicals can cause cancer, but it seems likely that they are not the major cause. To select an arbitrary figure, perhaps 5% of all cancers are caused by chemicals. Other known causes include radiation, viruses, chronic mechanical or thermal trauma, aging, parasitic infections, genetic predisposition, and perhaps even normal or accelerated cell division (mitosis). In most cases, however, the cause is unknown.

Not only is the cause unknown, but the fundamental way in which a normal cell is transformed into a cancer cell is also largely unknown. It is known that there are special kinds of genes in cell nuclei that are called proto-oncogenes. Although their usual function is not known with certainty they seem to have something to do with cell growth and division. When the cell is damaged in a special way, the proto-oncogenes are converted to active oncogenes. It is likely that there are numerous and complicated subsequent steps, some of which are irreversible, but the end result is that a normal cell has become a cancer cell. The process in which the transformed cancer cell evolves into a distinct tumor is regarded as a separate sequence of events.

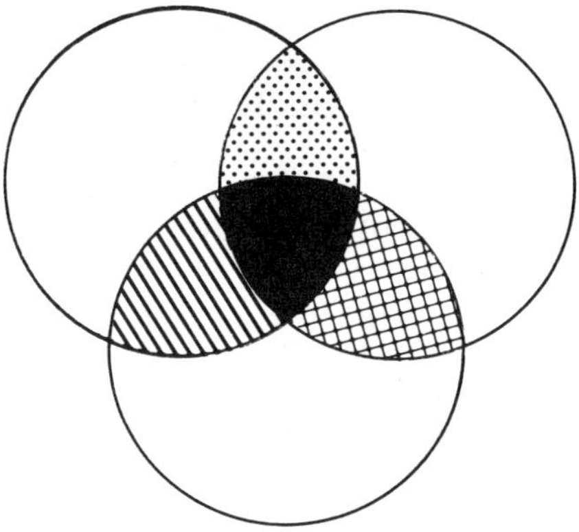

Figure 20–1. A symbolic representation of the overlap among chemicals that can cause cancer, mutations, and teratogenic effects. Areas of overlap show some chemicals produce all three; others produce only two of the three. Some chemicals produce only one of the three, but most chemicals produce none of the three.

Because cancer cells arise from normal cells, they retain most of the characteristics of normal cells and usually differ from them only in their tendency for unrestrained growth. This makes cancer cells much more difficult to kill than bacterial cells, which have different biochemical properties from normal human cells. Thus, bacteria are vulnerable to selectively toxic chemicals, antibiotics, but most anticancer drugs destroy some normal cells along with the cancer cells. Indeed, this lack of selectivity is limiting on the doses of anticancer drugs that can be given.

Although the growth of cancer cells is unrestrained, they need the same nutrients from the blood as normal cells, and in the case of solid tumors, growth may exceed the ability of the blood to bring new nutrients. Such tumors may have a center of dead and necrotic cancer cells. Some cancer cells grow faster than normal cells, but others grow more slowly. It is not the rate of growth that is characteristic, but the property that allows them to continue to grow beyond the normal needs of cellular repair or replacement. Certain forms of cancer are manifested as solid tumors that can be removed surgically, but other forms of cancer are disseminated (widely scattered in the body), such as the leukemias, which are characterized by an over production of white blood cells. Some cancer tumors are more or less benign, and others are classified as malignant. Malignant cancers are characterized by their ability to invade surrounding normal tissue and to metastasize to distant sites in the body. In metastatic cancers, cells have broken off from the primary focus, and traveled in the blood to some other site in the body and started growing again as secondary tumors.

Cancer is the second leading cause of death in the U.S. after coronary artery heart disease. About one in every four Americans will die from some form of cancer. Naturally, there is a great deal of concern about such a common, and often fatal disease. Because the cause of most cancers is unknown, attention has focused on chemicals, which are one of the known causes, and enormous effort has gone into the regulation of known or suspected carcinogens. An immediate problem is the recognition of cancer-causing chemicals. According to Dr. David P. Rall, the former Director of the National Institute of Environmental Health Sciences, approximately 26 chemicals have been firmly associated as causes of human cancer (Table 20–1).

The chemicals or industrial processes in Table 20–1 were identified at least in part by epidemiologic studies, which is the least ambiguous basis on which to regulate them. The first crude epidemiologic study may have been in 1775 when Dr. Percival Pott

TABLE 20–1. CHEMICALS KNOWN TO BE CARCINOGENIC IN MAN

Aflatoxins
4-Aminobiphenyl
Arsenic compounds
Asbestos
Auramine (manufacture of)
Benzene
Benzidine
Bis(chloromethyl)ether
Cadmium oxide
Chloramphenicol
Chromium (chromate-producing industries)
Cyclophosphamide
Diethylstilbesterol
Haematite (mining)
Isopropyl oil
Melphalan
Mustard gas
2-Naphthylamine
Nickel (nickel refining)
N,N-Bis(2-chloroethyl)-2-naphthylamine
Oxymetholone
Phenacetin
Phenytoin
Soot, tars, and oils
Vinyl chloride

(After Maugh TH II: Chemical carcinogens: the scientific basis for regulation. Science 201:1200, 1978.)

of London pointed out an increased incidence of cancer of the scrotom among chimney sweeps. Much more recent examples of cancer-causing chemicals that were identified by epidemiologic studies are benzene, asbestos, and arsenic, but in a practical sense, epidemiology is not much help to the regulatory agencies. As indicated in Chapter 3 cancer-causing chemicals have a uniquely prolonged latent period between first exposure and the appearance of the effect. On average, that latent period is 20 years. By the time a new chemical has been identified by an epidemiologic study as a carcinogen, a large number of people already may have been exposed. Moreover, possible subjects for inclusion in the study may have already died from other causes, moved out of the area, or have been exposed to other carcinogens or hazardous chemicals. One usually cannot rely on epidemiology alone to prove whether or not a chemical is carcinogenic, but benzene was identified as a human carcinogen before it was shown to cause cancer in animals. The best epidemiologic studies involve clusters of cases of a relatively rare form of cancer among groups of people who have much in common, such as occupation or place of residence. Thus, the chemicals in Table 20–1 are mostly industrial chemicals, and the cancers occurred among people who worked with them.

The alternative to epidemiologic studies is animal testing, which is based on the assumption that chemicals that produce cancer in rats will also produce cancers in humans. The accuracy of that assumption has been hotly debated, and the studies are costly. One large contract house prices a two-year oncogenicity study in rats at $585,000. Perhaps in time studies in whole animals will be replaced by less costly and shorter term tests in bacteria and cultured cells, but this places an even greater strain on the credulity of the assumption that the results of these tests have relevance for human cancer. One estimate of the total number of chemicals that have had adequate oncogenicity studies in animals is 3500 and of those about 750 were judged to be carcinogenic.

There are several different ways in which an oncogenicity test can be judged as having yielded positive results. As animals or humans age, they tend to develop tumors "spontaneously," and this must be taken into account in animal testing. Thus, the appearance of a tumor type in the exposed group that is not found in the control group to a statistically significant degree is one kind of positive result. A second kind of positive result would be the finding of the same kind of tumor in both exposed and control groups, but with statistically significant increases in the incidence (the percentage of the animals affected) in the exposed group. The third positive result would be a statistically significant decrease in the latency to the appearance of tumors in the exposed group. The fourth kind of positive result can occur when the incidence is the same in both control and exposed groups but the numbers of tumors per animal (multiplicity) is increased in the exposed group.

Carcinogens have all the characteristics of other toxic chemicals in that they show a dose-response relationship, they can undergo biotransformation, and their effects vary with all the factors discussed in Chapter 3. As indicated in Chapter 4, however, they are the example par excellence of hit-and-run poisons with a high tendency to produce irreversible effects. A direct-acting carcinogen is one that reacts chemically with DNA, and it would do so in a test tube as well as in an animal. The cell has several mechanisms by which it can repair damage to its DNA, but if an appropriate irreparable lesion is produced, the damaged DNA may activate a proto-oncogene to start the sequence of transformation to a cancer cell. These chemicals are also said to be genotoxic and mutagenic. A procarcinogen is a chemical that must be biotransformed into a direct-acting carcinogen. Paradoxically, many procarcinogens and direct-acting carcinogens are found among the drugs that are used to treat cancer. One way to kill cancer cells is to damage their DNA but, as already indicated, the drugs are not specific. Because of drug-induced DNA damage in normal cells, secondary cancers may arise in patients years after they have been cured of their primary disease. Asbestos is a so-called epigenetic carcinogen because it does not react chemically with DNA, and it does not undergo biotransformation to a direct-acting carcinogen. Epigenetic carcinogens are the most mysterious of all because even less is known about how they exert their effect, but they must be cytotoxic by some other mechanism.

In conducting oncogenicity testing it is common to feed or administer large doses of the chemical. Often the maximum tolerated dose (MTD), or a dose just short of the lethal dose because of other toxic effects for that test species is given. In many cases the doses are hugely unrealistic in terms of the amount of chemical that might be absorbed under conditions of ordinary environmental exposures. This is done to minimize the possibility of a false negative result in which a clean bill of health might be given to a chemical that is actually carcinogenic. Unfortunately, this design maximizes the chances for false positive results in which potentially useful chemicals may be banned or strictly regulated. Another potential complication occurs in the case of procarcinogens. It is possible that the doses are so high that the metabolic pathway for their activation becomes saturated, and that increased dose fails to show increased oncogenic effect. The risks for humans could then be underestimated at higher dosing levels.

The modern view of carcinogenesis is that it occurs in two stages. The first stage of initiation involves exposure to a pro- or direct-acting carcinogen. This results in transformation of a normal cell to a cancer cell. The second stage involves promotion of growth of cancer cells into actual tumors. Promoters can be other chemicals, but other factors probably promote tumor growth as well. In order for carcinogenesis to occur, the exposure sequence must be in the order of initiator then promoter. Promoters given before the initiator have no effect. The initiator can be effective in a single dose, but divided doses are more effective in some cases. The promoter needs multiple doses for maximal activity. When given alone, promoters can sometimes increase the incidence of spontaneous tumors, and in that sense they are epigenetic carcinogens. "Complete" carcinogens act as both initiators and promoters.

Because of the enormous expense associated with testing for carcinogenic activity, toxicologists have experimented for years with short-term, in vitro tests just as they have developed 90-day studies as possible substitutes for chronic toxicity studies (see Chapter 5). Carcinogen testing is more sophisticated, however, and the tests are arranged in an order of hierarchy with decision junctures at various points about whether or not a full scale, long-term, in vivo carcinogenicity test is indicated or not.

Even before actual testing begins, a study of the chemical structure of the compound in question may provide hints about its biologic activity or about how it might undergo biotransformation to an active carcinogen. The structural resemblance of dimethylhydrazine to established carcinogens first led to the suspicion that it might also induce cancer. The polycyclic, aromatic hydrocarbons have been extensively studied because some, such as benzo[a]pyrene, are well known carcinogens. Some others in the group were devoid of activity, however, and the molecular details of the requirements for that activity are now well defined.

One of the short-term tests used for carcinogenicity is also a test for mutagenicity using bacteria. It is based on the considerable evidence that DNA is the critical target for most carcinogens, and on the observation that mutagenic chemicals are often carcinogenic as well. The final leap of faith is that the test results have relevance for human cancer. This is the test that was developed by Dr. Bruce Ames and his colleagues and has since been refined in many variations. One variation uses a strain of bacteria that depends completely on the amino acid, histidine, which must be included in their media for growth and survival. These bacteria are actually mutants from an original strain that was histidine-independent. The bacteria are incubated with a direct-acting mutagen/carcinogen in a media containing histidine. After an appropriate period of time, the bacteria are transferred to a medium devoid of histidine, but still containing all other nutrients required for growth. If nothing happened, the bacteria will not survive, but if the bacteria underwent a so-called back mutation, it is apt to be one that permits them to survive again in the absence of histidine. The chemical can then be said to be mutagenic.

As indicated above, a large group of chemicals are procarcinogens, and must be biotransformed in order to become direct-acting. Bacteria lack the P450 isozyme systems that activate procarcinogens in animals and man. A modification of the Ames test, which permits it to detect some procarcinogens, involves a preincubation of the procarcinogen with mammalian liver homogenates, microsomes, or purified and isolated P450 systems to provide the activation step and the bacteria to provide the detection step. Other modifications that increase the sensitivity include selection of bacterial strains that have limited ability to carry out DNA repairs and selection of strains that have cell wall defects to make them more permeable to test chemicals or their biotransformation products.

Continuously reproducing mammalian cell culture systems permit modifications of the Ames test for mutagenicity to be carried out in cells that more closely resemble their human counterparts. Cell cultures can also be primary, such as fresh cells obtained from an excised organ, which are used only for short-term tests in which it is not necessary for the cells to reproduce. There are ways to measure the extent of DNA repair in primary cell cultures, and this is another type of short-term carcinogenesis bioassay. Eukaryotic cells undergo mitosis or cell division, whereas prokaryotic cells like bacteria do not. During the phase of mitosis in the cell lifecycle the chromosomes containing DNA are arranged in the shape of Xs with a so-called centromere at the "crossing" point. At this phase they are highly compacted, and can be seen in an ordinary microscope.

The normal human has 46 individual chromosomes in the set called the karyotype. These are arranged in 22 pairs of autosomes and one or the other of the two paired sex chromosomes. The Xs are made up of two chromatids, and when they are on the same side of the centromere, they are called sister chromatids. One set of arms is usually longer than the other set. Certain kinds of damage to the cell or its nucleus lead to an exchange of sister chromatids in which they switch sides without any other change in their shape or appearance. A special dye allows one to evaluate the extent of sister chromatid exchange, which is believed to be the most sensitive test yet devised for genetic damage. Finally, tests for actual transformation of normal cells to cancer cells obviously have great relevancy because it has been shown that for these systems a correlation exists between transformation and actual malignant tumor growth. The latter is not part of the test as it is actually performed.

In vitro and in vivo short-term tests are also done for carcinogenic promoting activity because it is important for regulatory affairs to distinguish between a promoter and a carcinogen. The third level of hierarchy involves short-term in vivo bioassay systems (40 weeks or less) including the induction of skin, lung, or breast cancers in mice. Much interest has centered around so-called preneoplastic lesions in rodents. These lesions can be distinguished readily on microscopic examination of tissue sections, but they are not yet true cancers. It is the premise, that given sufficient time most or all of these preneoplastic lesions would go on to become actual tumors. This test can yield meaningful results within 3 months so that it becomes truly analogous to 90-day toxicity studies.

The final stage would be that of a chronic bioassay. The advantage of such testing is that it is the most relevant because it shows the actual development of cancer in animals; the disadvantage is that all the uncertainties about extrapolation from animals to humans still exist. Other disadvantages have already been alluded to in terms of high costs and the long periods of time needed.

The conventional wisdom of chemical carcinogenesis has been challenged by a meticulous nonprofessional, Edith Efron (1984), and by Dr. Bruce Ames (Ames and Gold, 1990). Efron attacked the scientific foundations for the apocalyptic predictions of endemic cancer caused by synthetic industrial chemicals, pollution, and pesticide residues, and showed that little or no scientific basis exists for such assertions. Ames has placed his international reputation on the line to stand with her. Some of Ames' arguments are that humans are exposed to natural carcinogens in food in numbers and at levels that are orders of magnitude above strictly regulated synthetic chemicals. The natural carcinogens and oxidative products of normal metabolism cause more frequent damage to DNA than previously appreciated. Those lesions are converted to stable mutations by cell division. Thus, a key argument is that induced cell division (mitogenesis) plays an important, if not dominant, role in carcinogenesis. Ames believes that rodent testing at the MTD maximizes the chances of positive findings because it tends to increase the rate of mitogenesis and nonspecific inflammation rather than increasing the damage to DNA, which might occur also at lower doses. More than half of all synthetic and naturally occurring chemicals are carcinogenic in the MTD test in rodents, and a high percentage of those are not mutagenic. Therefore, it is possible that mitogenesis may explain the phenomena of promotion and epigenetic carcinogenesis. Whether or not these hypotheses will stand the test of time remains to be seen, but they have certainly stimulated some novel thinking in an old field.

MUTAGENESIS

There are three basic ways in which damage can be inflicted at the genetic level. The first, called aneuploidization, involves a gain or loss of one or more whole chromosomes. The second is called clastogenesis and involves the breaking off of a piece or pieces of a chromosome, which can result in the gain, loss, or rearrangement of those pieces. The third way is a point or gene locus mutation in which there are relatively small additions or deletions in the base pairs of the DNA sequence. As noted above, one possible ultimate manifestation of genetic damage is the transformation of a normal cell to a cancer cell. The germ cells of the body are the spermatazoa in males and the egg in females; all other types of cells are called somatic cells. The concern in genetic damage is with the germ cells because the damage can be passed on to subsequent generations. Damage to somatic cells is not passed on to the progeny, but it can result in cancer in that individual.

Out of each 1,000,000 human conceptions there are about 150,000 spontaneous abortions, and examination of the karyotype of the fetuses show that about half, or 75,000, had chromosomal abnormalities. Of the 850,000 live births, there are about 17,000 perinatal deaths occurring between the 28th week of gestation and the first 7 days after birth. Of the surviving 833,000 children about 5000 have chromosomal abnormalities. Thus, a certain amount of "genetic drift" is inevitable in human and animal populations, as recognized many years ago by Charles Darwin.

Point mutations can result in heritable genetic diseases, such as color blindness and sickle cell anemia, certain congenital malformations and retardations, and other conditions with variable age of onset, such as cystic fibrosis. It is estimated that these conditions in toto affect about 1% of all newborn infants. Diseases associated with mutagenesis increase in frequency with increasing paternal age.

The occurrence of clastogenesis is detected by cytogenetic analysis in which the karyotype of stained chromosomes is examined microscopically. It is detected in about 0.25% of all newborns, but the frequency increases with increasing maternal age. It is widely believed that clastogenesis results in a greater tendency for spontaneous abortion, i.e., it creates an intrauterine selective pressure to abort. Aneuploidization results in some of the more common inherited genetic diseases, such as Down's syndrome, and in somatic cells it has been seen in some forms of cancer. It also seems to exert a strong intrauterine selective pressure for spontaneous abortion.

Some of the ways in which bacterial systems can be exploited to test for mutagenesis were described above, but tests have also been devised using mammalian systems. One of these is the dominant, lethal assay, which usually uses rats or mice. A male animal is given a single dose of the suspected chemical, and then mated with two untreated females each week for eight weeks. The females are sacrificed before reaching term, and the number of live embryos and the sites where reabsorption of dead embryos has occurred are counted. A positive result will show an increase in the ratio of dead to live embryos.

There are numerous variations around the so-called host-mediated assay for mutagens. As indicated above a short coming of the tests involving bacteria is that the bacteria lack P450 isozyme systems for activating mutagens. Although the addition of a microsomal fraction (S9) from rodent livers to the system, partially compensates for this drawback, it does not completely mimic an intact liver. Moreover, there are other sites for xenobiotic biotransformation, such as the intestinal microflora, that are often lacking in the bacterial assay. The host-mediated assay takes advantage of all the xenobiotic biotransforming capabilities of an intact animal, usually a rodent, irrespective of its anatomic locus. It also includes not only activation reactions, but inactivation pathways as well. The animal is the host, and it is inoculated with any of a variety of indicator organisms, such as mammalian cells, fungal spores, bacteria, or yeasts.

The host is then given the test chemical. After an appropriate period of time the indicator organism must be recovered, and subjected to any of a number of appropriate tests for genetic damage, such as DNA repair. The indicator organism might be placed in the blood by intravenous injection and recovered by drawing a blood specimen. It is a highly useful and relevant test system both for studies on mutagenesis and carcinogenesis. A third mammalian test system is to treat the animal with the suspect chemical and then take samples of bone marrow and examine the karyotypes of the cells microscopically for evidence of chromosomal damage (in vivo cytogenetics).

TERATOGENICITY

Agents that produce defects in development and growth are called teratogens or mutagens. As indicated above, mutagens involve germ cell damage to DNA, and cause permanent mutations that result in a fetus that differs from the parents. Teratogens affect somatic cell lines and only the child with the affected cell line will differ from the parents. Malformed children were recognized before recorded history. Such births created intense feelings in different societies from horror and dread to sympathy and protection. A defect present at birth is said to be congenital, but it does not have to be a hereditary defect; those account for only about 20% of congenital defects. Unlike teratogenic effects, they may be passed on to subsequent generations. There are about 250,000 birth defects of all degrees of severity in the United States each year, and the causes for about two-thirds of these are unknown. Only about 3% can be ascribed to exposure to drugs or chemicals. Defects severe enough to require some form of treatment occur in 3 to 7% of human births.

The first teratogen to be recognized as such in 1941 was measles (rubella) virus. An epidemic of measles in Austria naturally included a number of pregnant women and the incidence of subsequent birth defects was elevated to the point where it could no longer escape notice. One of the important findings to emerge from these epidemiologic studies was that the nature of the birth defect depended on the stage of development of the fetus at which exposure occurred. If the infection occurred during the first or second month of gestation, defects were found in the heart or eyes. If the disease occurred during the third month, the defect involved hearing. The risks were highest during the first month of pregnancy. Infections after 3.5 months of pregnancy carried about a five-fold lower risk and involved hearing, speech, and mental deficits rather than physical malformations.

The insidious aspect of many teratogens is that the damage is done so early in pregnancy. It is not unusual for a woman to be pregnant for two months and not be aware of it. By the time the results of pregnancy testing are in hand it may already be too late to terminate or avoid exposure. Women infected with rubella are well aware that they are infected even though they may not know that they are pregnant, and prophylactic measures can be taken although they are by no means completely effective. Other teratogens do not cause symptoms in the mother, and she may not be aware either that she is exposed to a teratogen or that she is pregnant.

The full impact of the devastation that teratogenic agents can cause was brought forcefully to the public attention in 1961 when the association was made between severe malformations in infants and the ingestion of the drug, thalidomide, during the early months of pregnancy by the mothers. The drug had already been on the market for five years in what was then West Germany. Thalidomide was marketed as a sedative, but it was widely used to control the nausea and vomiting of morning sickness during early pregnancy. This unfortunate use led to the exposure of thousands of pregnant women during critical periods of their gestation. In 1961 over 150 children in West Germany were born with rare limb malformations. Affected children had varying degrees of phocomelia with reductions in length of the long bones of the limbs usually affecting the arms more than the legs, or amelia in which there was a complete absence of limbs. Many other kinds of defects occurred as well, but the limb defects were the hallmark of the pathetically crippled children. Eventually, cases were reported from England, Wales and, Japan, and estimates place the total number of affected children at 7000 to 8000. Because approval of thalidomide was delayed by the FDA, the U.S. population largely escaped the disaster.

The thalidomide tragedy stimulated intense interest and research in the area of teratogenesis and solidified it as a subspecialty of toxicology. It shifted the emphasis from inherited defects to those induced by chemicals during gestation. In attempts to find an animal model for the malformations it was discovered that the range of species sensitivities to the effect was widespread, with rodents being relatively resistant and rabbits relatively sensitive. Factors such as absorption, biotransformation, and distribution in the body and to the fetus have largely been ruled out as accounting for these species differences. A number of theories have been proposed to account for the basic mechanisms involved in the malformations, but none have been adequately proved or disproved. The mechanisms at work are still basically unknown.

What is known today is that despite large differences in the total length of gestation considerable similarity exists among species in the timing of the early events of pregnancy, especially the critical period of organogenesis when the rudimentary organs begin to develop in the fetus. Various organs develop at different but overlapping times in the embryo in an intricately programmed sequence. For example, in rats, development of the eyes is early, followed by heart, skeleton, and brain, whereas development of the urogenital system occurs much later. Thus, the differing patterns of malformations depend on the period of organogenesis at which exposure occurs as well as on the specific chemical. It is possible that teratogens damage a single critical cell or a small cluster of cells responsible for initiating the development of an organ causing the organ to develop abnormally or to fail to develop altogether.

For many years a central assumption has been that exposure to a teratogen must occur in the uterus of the pregnant mother. The placenta is the organ that forms in the lining of the uterus and joins the fetus to the mother through the umbilical cord, which provides the fetus with nutrients and removes its waste products. Although the placenta acts as a partial barrier to protect the fetus from foreign chemicals, it is a rather leaky barrier and unfortunately allows many teratogens to pass through. Some sketchy evidence is beginning to accumulate, however, to implicate the father in the genesis of some teratogenic events. The dominant lethal effect (above) is certainly a result of exposure of the male, but it seems possible that actual malformations or other developmental anomalies also may be passed on to the offspring by the father. Whether this could result from damage to the sperm or its DNA, or whether the chemical is actually present in the seminal fluid is not known. Despite many allegations there is no firm evidence that this has happened in humans, but animal experiments suggest that when male rats were exposed to lead, their offspring exhibited changes in brain development. There is certainly increasing concern that men exposed to lead, ethanol, some anticancer drugs, nuclear radiation, and dioxin-containing herbicides could be conceiving children with birth defects.

SUMMARY POINTS

Cancer, genetic damage, and birth defects are the most feared of all toxic effects; chemical agents, however, account for only a small number of the total cases.

Cancer is the second leading cause of death in the U.S., but only about 5% of all cases are caused by chemicals; less than 30 chemicals are known to cause human cancer.

Cancer cells are similar to normal cells; some grow faster than normal cells and some grow slower, but they lack the usual restraints to growth of normal cells.

Malignant cancers are characterized by their tendency to invade normal tissue and to metastasize to sites in the body distant from the original cancer.

Epidemiology is highly relevant, but of limited value because the latent period between exposure to a carcinogen and the appearance of cancer averages 20 years.

Animal testing may be more relevant, but it is costly and time consuming; short term bioassays, such as the Ames test, may be used to see if animal testing is justified.

Cancer is thought to occur in two stages; in the initiation stage there is permanent damage to DNA in normal cells and in the stage of promotion the cancer cells multiply.

Most carcinogens must undergo biotransformation to active agents that react with DNA, but epigenetic carcinogens do not appear to react with DNA.

SUMMARY POINTS (continued)

In bacterial tests for mutagenicity (one indication that a chemical may be a carcinogen) liver enzymes must be added to the system to carry out biotransformation.

Maximum tolerated dose (MTD) animal testing has been criticized recently because it may initiate mitogenesis, which increases the possibility of mutation to a cancer cell.

A mutation in a germ cell can be passed on to subsequent generations; a mutation in a somatic cell cannot be passed on, but it can result in cancer.

The dominant lethal assay for mutagenicity tests for effects in male reproduction; the host-mediated assay is more realistic because it is conducted in a whole animal.

Teratogens affect only somatic cell lines at the time of organogenesis so that the tissue, organ, or limb develops abnormally or fails to develop altogether.

Of the 250,000 birth defects each year in the U.S., 3% are thought to be due to chemicals; exposure may occur in women who do not know that they are pregnant.

If the teratogen does not produce signs of toxicity in the mother, she may not be aware that she was exposed.

BIBLIOGRAPHY

Ames BN, Gold LS: Too many rodent carcinogens: mitogenesis increases mutagenesis. Science 249:970, 1990.

Efron E: The Apocalyptics. New York, Simon and Schuster, 1984.

Gifford RW Jr: Effects of toxic chemicals on the reproductive system. JAMA 253:3431, 1985.

Maugh TH II: Chemical carcinogens: the scientific basis for regulation. Science 201:1200, 1978.

Maugh, TH II: Chemical carcinogens: how dangerous are low doses? Science 202:37, 1978.

Rao KS, Schwetz BA, Park CN: Reproductive toxicity risk assessment of chemicals. Vet Human Toxicol 23:167, 1981.

Shubik P: Criteria for evidence of chemical carcinogenicity. Science 225:682, 1984.

INDEX

Page numbers in italics refer to figures. Page numbers followed by "t" refer to tables.

Abortion(s), spontaneous, 281, 282
ABS. *See* Alkyl benzene sulfonate(s) (ABS)
Absorbed sulfite(s), 166
Absorption, of ethanol, 262–263
experimental techniques for injecting chemicals, 66
through gastrointestinal tract, 70–73
of heavy metals, 143, 147–148
of lead, 147–148
through peritoneal cavity, 66
through respiratory system, 66–70, *67*, 69t
through skin, 73–74
through veins, 66
Abstinence syndrome, 269, *270*
Acceptable daily intake (ADI), 94t, 95
Acetaminophen, 25
Acetone, 246
Acetylcholine (ACh), 7–8, 109
Acetylcholinesterase (AChE), 109
ACGIH. *See* American Conference of Governmental Industrial Hygienists (ACGIH)
ACh. *See* Acetylcholine (ACh)
AChE. *See* Acetylcholinesterase (AChE)
Acid rain, 167–168
Acidosis, 256
Acquired tolerance, 267–268, *268*
Activated charcoal, 72
Active transport, 64
Acute effect, 15
Acute toxicity, chronic toxicity compared with, 36
distribution of sensitivities and, 15–18, *16–18*
factors affecting, 22–24
latent period and, 24–25
minimal lethal dose (MLD), 15
relationship with chronic toxicity, 45–46, 45t, 47
Acute toxicity testing, for pesticides, 53–56, 54t, 55t
Additive toxicity, 26, 29
Additive(s) in food. *See* Food additive(s)
ADI. *See* Acceptable daily intake (ADI)
Adsorption, 65–66, *65*, 74
Aerosol(s), 69, 69t, 163–165, 167–168
Agency for Toxic Substances and Disease Registry (ATSDR), 239, 239t
Agent Orange, 22, 134, 138
Agriculture, Department of, 84, 86, 93
Air pollution, absorbed sulfites, 166
acid rain, 167–168
chemicals in urban air pollution, 183
chronic toxicity testing and, 43–44
hydrogen sulfide, 157–159
increases in human death rate and, 160–162, *161*
indoor air pollution, 210–216
oxidizing type of, 183–184
sulfur and particulates, 160–168
sulfur cycle and, 153–160, *154*, 156t
sulfur dioxide, 157
woodsmoke, 165–166
Albumin, 144t
Alcohol(s), absorption of, 262–263
abstinence syndrome for, 269, *270*
biotransformation of, 263–265, 270
denatured alcohol, 258–259
distribution throughout body, 263
ethanol, 257, *257*, 258–271, *261*, *267–271*
ethanol-disulfiram reaction, 265

Alcohol(s) (*Continued*)
ethylene glycol, 257–258, *257*, 259, 262
excretion of, 263
fermentation of, 259–260
impact of temperature on, 24
inhalation of, 261–262
inhibitors of, 264–265
isopropanol, *257*, 258, 259
methanol, 256–257, *257*, 262
neuronal membrane fluidity hypothesis and, 268–269, *269*
physical and chemical properties of, 256
proof system for expressing alcohol content, 260–261, *261*
signs of intoxication, 266–267, *267*
subjective effects of alcoholic beverages, 265–266
tolerance to ethanol, 46–47, 267–270, *268*, *271*
use of alcoholic beverages, 265
Alcohol ethoxylate, 190
Alcoholic(s), definition of, 265
Alcoholic beverages, 258–271, *261*, *267–271*
Aldehyde(s), 183
Aldehyde dehydrogenase(s), 264
Aldicarb, 109, 110
Aldrin, 89t, 120
Alexander VI, Pope, 3
Algae, 197–200, 206
Alkyl benzene sulfonate(s) (ABS), 191, *192*, 206
"Alkyl" chain(s), 189, 191
All-or-none (quantal) response, 15
Allergen(s), 13, 70, 211, 222–223
Alveolar region, 67–69, *67*
Amdur, Mary, 163
American Conference of Governmental Industrial Hygienists (ACGIH), 41, 42, 158
Ames, Bruce, 279, 281
Ames test, 279–280
Amine(s), 182
Ammonia, 215
Ammonia gas (NH_3), 173, 176
d-amphetamine, 222
Amygdalin, 24
Anaerobic bacteria, 99–100
Anemia, 149
Anesthetic(s), 270
Aneuploidization, 282
Anionic surfactant(s), 190, 193, 206
Antabuse, 258
Antidote(s), 26–29
Antiepileptic drug(s), 80
Antifungal drug(s), 98
Aplastic anemia, 122
Aromatic(s), 129
Arsenic, 3, 57, 87, 101, 243, 277
Artificial color(s). *See* Color additive(s)
Asbestos, 43, 70, 211, 215, 238, 241, 277, 278
Aspirin, 75, 220
Atmospheric pollution. *See* Air pollution
Atrazine, 105, 106
Atropine, 109
ATSDR. *See* Agency for Toxic Substances and Disease Registry (ATSDR)
Auxin(s), 104, *105*
Avicide(s), 107
Avon's Skin-So-Soft, 107
Azide, 27–29, *27*

BAC. *See* Blood alcohol concentration (BAC)
Bacillus thuringiensis, 111
Baking soda, 75
BAL, *145*
Barbiturate(s), 47, 121
BCF. *See* Bioconcentration factor (BCF)
Beer, 259–260
Behavior disorder(s). *See* Childhood behavior disorder(s)
Bentonite, 25
Benzene, 211, 240–241, 277
Benzene hexachloride, 89t
Benzene ring(s), 129–130, *129*
Benzene sulfonate(s), 191–193, *192*
Benzidine, 238
Benzpyrene, 80
Bernard, Claude, 4–8
BHA. *See* Butylated hydroxyanisole (BHA)
BHT. *See* Butylated hydroxytoluene (BHT)
Bile, *70*, 71, 74–75, 148
Biliary system, 74
Bilirubin, 75
Bioconcentration factor (BCF), 242–243
Biologic membrane(s), chemistry of, 60–61, *61*
functions of, 60
properties affecting transportation of chemicals across, 61–66, *62*, *64*, *65*
Biotransformation, 74–75, 78–82, 108, 181, 257–258, 263–265, 270, 283
Birth defect(s), 137

Black death, 101
Black lung, 67, 70
Blood alcohol concentration (BAC), 74, 262, 263–264, 267, 270
Blood-brain barrier, 143
Blue baby syndrome, 178–179
Bock beer, 259
Bone, 144, 148
Bone marrow, 148–149
Borgia, Cesare, 3
Borgia, Lucretia, 3
Brain, 143, 148
Breast milk, 75, 100, 121, 137
Breathalyzer, 74, 263
Bromine, 114
Bronchi and bronchioles, 67–68, *67*
Brown lung, 67
Brownian movement, 63
Bubonic plague, 101, 115
n-Butyl ester(s), 134
Butylated hydroxyanisole (BHA), 221
Butylated hydroxytoluene (BHT), 221
Byssinosis, 67, 70

Cadmium, 36, 142, 149, 241
Calcium, 57, 142, 144, 147, 148, 194, 201
Cancer, 25, 70, 81, 137, 145, 214, 215, 240, 274–281, *275*, 282
Cancer cell(s), 274–275
Carbamate type AChE inhibitors, 109
Carbaryl, 109
Carbon, 57, 198
Carbon dioxide (CO_2), 42, 43t, 105, 211, 216
Carbon disulfide, 43, 159
Carbon monoxide (CO), 4–6, 12, 43t, 68, 74, 79, 183, 211, 216
Carbon tetrachloride, 42, 43t
Carbonate(s), 201
Carbonyl sulfide, 159
Carboxyl group, 189, *189*
Carcinogen(s), 46, 89, 90, 106, 165, 181, 241, 250–251, 276t, 277–281
Carcinogenesis, 274–281, *275*
Carmoisine, 228
Carroll, Lewis, 4
Carson, Rachel, 125–126
Cationic surfactant(s), 190
Ceiling value(s) (CV), 42–43, 43t
Centromere, 280
CERCLA. *See* Comprehensive Environmental Response, Compensation, and Liability Act (CERCLA)
CH_3-HgCl. *See* Methyl mercuric chloride (CH_3-HgCl)
Charcoal, activated, 72
Chelating agent(s), *145*, 146
Chelation, 196
Chemical interaction(s), 26–32
Chick edema factor, 133, 135
Childhood behavior disorder(s), and animal experiments, 229–231
  clinical studies of, 223–229
  facts concerning, 222–223
  Feingold hypothesis for, 220–222
  food additives and, 220–231
Chitin, 111
Chlordane, 122, 211
Chloride(s) (Cl), 144t, 241
Chlorine (Cl_2), 39, 57, 114, 133
Chloroacne, 130
Chloroform, 23, 211, 241, 243
Chlorophenol(s), 130, 136t
Chromate, 145
Chromatid(s), 280
Chromium, 145, 241
Chromosome(s), 280, 282
Chronic toxicity, relationship with acute toxicity, 45–46, 45t, 47
  testing for, 43–47
  testing for carcinogenic effects, 50
Chronicity index (CI), 44–46, 45t
Chrysanthemum family, 108
CI. *See* Chronicity index (CI)
Cigarette smoking, 70, 80, 162, 183, 211, 214, 215, 216
Cimetidine, 80
Cirrhosis, 47, 81
Citronella, 107
Cl. *See* Chloride (Cl)
Cl_2. *See* Chlorine (Cl_2)
Clark, Mark, 114
Clastogenesis, 282
Clay(s), 25
Clean Water Act, 238
Cleopatra, 2
Clostridium botulinum, 26, 181
CO. *See* Carbon monoxide (CO)
CO_2. *See* Carbon dioxide (CO_2)
Coal, 157, 241
Colchicine, 2–3
Color additive(s), 89–90, 92, 220, 221, 223–224, 229–231
Color Additives Amendment, 89–90
Color blindness, 282
Compound 1080, 22
Comprehensive Environmental Response, Compensation, and Liability Act (CERCLA), 237
Confidence limits (CL), 32–33
Conjugation reaction(s), 81
Contact insecticide(s), 111

Copper, 56–57, *57*, 87, 142, 143, 149
Corn, Morton, 163
Cresol, 43t
"Cross-tolerance," 47
Cube, 108
Curare, 6–8
Curing, of meat and fish, 180–181
Cyanide(s), 24, 26, 27–29, *27*, 31–33, 45, 101, 238, 241, 242, 243
Cycasin, 182
Cysteine, 144t
Cystic fibrosis, 282
Cytochrome, 79
Cytochrome oxidase, 159
Cytochrome P450, 136, 145, 264

2,4-D, 105–106, 134
Darwin, Charles, 103, 282
DBF. *See* Dibenzofuran(s) (DBF)
DDE, 116–117, 119–120, *119*
DDT, ability to cross the human female placenta, 121
  as agricultural insecticide, 115–116, *117*
  aplastic anemia and, 122
  banning on ecological grounds, 122–126
  barbiturate drugs and, 121–122
  bioconcentration factor of, 243
  body fat concentrations of, 116–121, *119*
  compared with PCB, 131–133
  concentration in natural food chains, 123–125, *123*, *124*
  control of insect vectors in diseases, 114–115
  discovery of, 114
  insecticides related to, 45, 80, 89
  introduction of, 128
  measurement of, *92*
  physical and chemical properties of, 115
  tolerances and action levels of, 89t, 91
  USPHS surveys and, 117–118
Death rate, and acute air pollution, 160–162, *161*
Delaney clause, 89–92, 106
Denatured alcohol, 258–259
Denitrification, 174
Department of Agriculture, 84, 86, 93
Department of the Interior, 86
Dermatitis, 130
Dermis, 73–74, *73*
DES. *See* Diethylstilbesterol (DES)
Detergent builder(s), 193–197, 194t, 195t
Detergent controversy, current status of, 205–207
  detergent builders, 193–197, 194t, 195t
  effectiveness and safety of polyphosphates, 193–197, 194t, 195t
  efficacy and safety of nonphosphate builders, 202–205, *203*, *204*
  eutrophication, 197–200
  historical survey of soap and other surfactants, 188–193, *189*, *192*
  industry, government, and politics, 200–202
  phosphate-free detergents, 200–202
Diazepam, 269
Dibenzofuran(s) (DBF), 135, *135*, 136t
p-Dichlorobenzene, 216
Dicoumarol, 102
Dieldrin, 89t, 120, 121
Diethylene glycol, 87–88
Diethylstilbesterol (DES), 23
Diffusion, 63, 261
Dilan, 120
Dimercaprol, *145*
Dimethyl disulfide, 160
Dimethyl sulfoxide (DMSO), 156, 159
Dimethylnitrosamine (DMNA), 181–182
Dimethylsulfide (DMS), 155–156, 159, 160
Dinitrophenol, 24
Dinoseb, 86
Dioxin(s), 80, 135–137, *135*, 136t, 138
Direct-acting carcinogen(s), 278
Distribution of sensitivities, acute toxicity, 15–18, *16–18*
Disulfiram, 265
DMNA. *See* Dimethylnitrosamine (DMNA)
DMS. *See* Dimethylsulfide (DMS)
DMSO. *See* Dimethyl sulfoxide (DMSO)
DNA, 145, 278, 279, 280, 281, 282, 283, 284, 286
Documentation of the Treshold Limit Values, 41
Dominant, lethal assay, 283
Dose, relationship to toxicity, 14
  in terms of molecular weights, 13–14, 14t
Dose-mortality relationship(s), 27–29, *27*
Dose-response relationship, in acute toxicity, 15–20, *16–18*
  definition of, 13

representation on probit graph paper, 18–19, *18*
Dowicide(s), 130, *131*
Down's syndrome, 282
Drug-drug interactions, 81

ED50, and acute toxicity, 17–20
determination of, 19–20
limitations and uses of, 22–33
EDB. *See* Ethylene dibromide (EDB)
Edwards, Charles, 202
Effective dose 50. *See* ED50
Efron, Edith, 281
Egyptians, 2–3
Ehrlich, Paul, 3–4
Elixir of Sulfanilamide, 87–88
Encephalopathy(ies), 148
Endoplasmic reticulum (ER), 79, 81
Endrin, 89t, 120, 125
Enterohepatic recycling, 72
Environmental pollution. *See* Air pollution
Environmental Protection Agency (EPA), 53–56, 84–86, 89, 93, 95, 126, 158, 167, 202, 212, 214, 215
Environmental toxicology, historic insights on, 2–8
Enzyme(s), 79–80, 121–122, 146, 264
EPA. *See* Environmental Protection Agency (EPA)
Epidermis, 73–74, *73*
Epigenetic carcinogen(s), 278
Epiglottis, *70*, 71
ER. *See* Endoplasmic reticulum (ER)
Erythrosin B, 229–231
Ester(s), 134
Ethanol, 74, 80, 257, *257*, 258–271, *261*, *267–271*
Ethanol-disulfiram reaction, 265
Ethyl alcohol, 42, 43t, 46
Ethylene dibromide (EDB), 86
Ethylene glycol, 257–258, *257*, 259, 262
Eutrophication, 197–200
Excretion, 66, 74–75, 148, 263
Exposure potential, evaluation of, 248–249

Facilitated diffusion, 63–64
Farmer's lung, 67, 70
FDA. *See* Food and Drug Administration (FDA)
FD&C Yellow No. 5, 220
Feces, 74, 148
Federal Environmental Pesticide Control Act (FEPCA), 84–85, 89
Federal Food, Drug, and Cosmetic Act (FFDCA), 87, 88
Federal Insecticide, Fungicide, and Rodenticide Act (FIFRA), 84
Feingold, Benjamin, 220
Feingold diet, 223–229
Feingold hypothesis, 220–223
FEPCA. *See* Federal Environmental Pesticide Control Act (FEPCA)
Fermentation, 259–260
Fertilizer(s), 153, 175–176
FFDCA. *See* Federal Food, Drug, and Cosmetic Act (FFDCA)
FIFRA. *See* Federal Insecticide, Fungicide, and Rodenticide Act (FIFRA)
Filtration, 64–65, *65*
Firemaster, 134
Fixation, 172, *173*, 174–175
Fixed nitrogen, and fertilizers, 175–176
Fluorine, 114
Food additive(s), animal experiments on, 229–231
childhood behavior disorders and, 220–231
clinical studies of, 223–229
facts concerning, 222–223
Feingold hypothesis and, 220–222
laws on, 88, 89–90, 92
sulfite salts as, 166
Food Additives Amendment, 88
Food allergy, 222–223
Food and Drug Administration (FDA), 53, 87, 88, 89, 93, 95, 106, 159, 181, 202, 285
Food dyes. *See* Color additive(s)
Food law(s)
carcinogens, 89–90
color additives, 89–90, 92
food additives, 88, 89–90
glossary of food safety terms, acronyms, and standard values, 94t
history of, 86–88
setting tolerances for, 84, 88–90, 89t, 93–96
zero tolerance and, 90–92
Formaldehyde, 39, 42, 43t, 70, 211, 213–214
Frasch process, 152
Fumigant(s), 98
Fungal infection(s), 98
Fungicide(s), 98–101

Gas(es). *See* Inhalation toxicology; and names of specific gases
Gas chromatography, 120
Gas-liquid chromatography (GLC), 91

Gastrointestinal tract, absorption of ethanol, 262–263
absorption of heavy metals, 143
absorption of lead, 147–148
absorption through, 70–73
anatomy of, 70–71, *70*
excretion and, 74–75
Gaussian distribution, 16
Gene locus mutation, 282
Genetic disease(s), 282
GLC. *See* Gas-liquid chromatography (GLC)
GLC detector systems, 91
Gluthathione, 145
Glycoside, 24
Graphite, 25
GRAS status, 88, 159, 166
Greeks, 3
Groundwater contamination, 243–244

H_2S. *See* Hydrogen sulfide (H_2S)
H_2SO_3. *See* Sulfurous acid (H_2SO_3)
H_6SO_4. *See* Sulfuric acid (H_2SO_4)
Haber process, 172
Hair, 75
Halogenated aliphatic hydrocarbon(s), 238, 239t
Halogenated hydrocarbon(s), 212
Halogenated hydrocarbon insecticide(s) (HHI), 114, 115, 120, 121, 122, 132, 138
Halothane, 270
Halowax, 128
Hazardous, compared with toxic, 12
Hazardous waste site(s), benzene, 240–241
chloroform, 241
evaluation of exposure potential, 248–249
groundwater contamination, 243–244
hazardous waste superfund, 236–237
human contact with waste constituents, 242–243
human impact assessment, 250–251
identification of potential migration and fate of waste constituents, 247–248
lead, 240
legislation concerning, 236–237, 252
Love Canal, 234–235
national priority list of waste sites, 237–238, 238t
past waste disposal practices, 235–236
priority pollutants, 238–242
remedial action for, 251–253
risk assessment process, 244–251
site description, 245–247
toluene, 240
toxicological evaluation, 249
trichloroethylene, 239–240
HCB. *See* Hexachlorobenzene (HCB)
HCN. *See* Hydrogen cyanide (HCN)
Heavy metal(s), biologic effects of pairs of metals, 142–143
biotransformation of, 144–145
definition of, 142
diagnosis of metal poisoning, 146
ingestion of, 143
inhalation of, 143
ionic forms of, 143–144
ligands and, 143–144, 144t, 145–146
metal storage in the body, 144
prevention of metal poisoning, 146–147
therapeutically useful chelating agents, 145, *145*
treatment of metal poisoning, 146
water solubility of, 143
Hemlock, 3
Hemodialysis, 75
Hemoglobin, 148–149, 177
Heptachlor/heptachlor expoide, 89t
Herbicide(s), 94–95, 103–107, 130, 134
Heritable genetic disease(s), 282
Heroin, 46–47, 268, 269
Heroin abstinence syndrome, 269
Hexachlorobenzene (HCB), 100
Hexachlorophene, 137
HHI. *See* Halogenated hydrocarbon insecticide(s) (HHI)
Histidine, 144t, 279
"Hit-and-run" poison(s), 46
HNO_3. *See* Nitric (HNO_3) acid(s)
HO. *See* Hydroxyl (HO)
Hornig, James, 165
Host-mediated assay for mutagens, 283
Human contact with hazardous waste constituents, 242–243
Human impact assessment, 250–251
Hydrocarbon(s), 183, 242
Hydrogen, 57
Hydrogen cyanide (HCN), 36, 39, 42, 43t
Hydrogen sulfide (H_2S), 36, *37*, 39, 40, 41, 42, 43t, 68, 152, 155, 157–159
Hydropericardium, 133
Hydrophilic, 60
Hydrophobic, 60
Hydrothermal vent, 158–159

Hydroxyl (HO), 144t
Hyperactivity. *See* Childhood behavior disorder(s)
Hyperkeratosis, 128, 130
Hyperkinesis. *See* Childhood behavior disorder(s)
Hyperpigmentation, 136

In vivo cytogenetics, 283
Incineration, 252
"Indemnification" rule, 85–86
Indoor air pollution, 210–216
Industrial waste. *See* Hazardous waste site(s)
Infantile methemoglobinemia, 178–179
Infectious microorganism(s), 211
Ingestion. *See* Gastrointestinal tract
Inhalation toxicology, adjustments of toxic gas concentrations in experiments, 40–41
  ceiling values (CV) and, 42–43, 43t
  chronic toxicity and, 43–47
  continuous exposure to toxic concentrations of hydrogen sulfide, 39
  evaluation of toxicity of gas, 37, *37*
  heavy metals, 143
  latent period and, 36–37
  LC50 and, 40–41
  measurement of exposure in terms of ambient concentration of gas, 38–39
  nitrite-treated animals and exposure to toxic concentrations of hydrogen sulfide, 39–40
  short-term exposure limits (STEL) and, 42–43, 43t
  special problems with, 36, 38, 43–44, 46–47
  threshold limit values (TLV) and, 41–43, 43t
Inorganic lead poisoning, 148
Insect repellant(s), 107
Insect sex attractant(s), 111
Insecticide(s), 107–111. *See also* DDT; Pesticide(s)
Interior, Department of, 86
Intoxication, 266–267, *267*
Intoximeter, 74, 263
Inversion, 161, 183
Iodine, 114
Iron, 142
Isobologram, 29–32, *30, 31*
Isopropanol, 258, 259
Isozyme(s), 79, 80, 81. *See also* P450 isozyme(s)
Itai-itai disease, 149–150

Johnson, Samuel, 105

Kanechlor, 133
Kaolin, 25
Karotype(s), 280, 283
Kelthane, 120
Kepone, 120
Keratin, 128
Kerosene, 115, 242
Kidney(s), 75, 148, 149, 258
Koop, C. Everette, 215

Labeling, 84–85
Laetrile, 24
Latent period, 24–25, 37
Lavoisier, Antoine, 172
Laws. *See* Food laws; Hazardous waste site(s); Pesticide laws
LC50, 40–41, 54, 54t
LD50, and acute toxicity, 16–19
  age and body weight and, 23
  ambient temperature and, 24
  chronicity index and, 44–45, 45t
  diets and, 23–24
  differences among species in, 22–23
  health status and, 23
  and interactions between two chemicals in body, 26–32
  isobolograms and, 29–32, *30, 31*
  latent period and, 24–25
  route of chemical exposure and, 24
  sex differences in, 23
  signs of toxicity and, 25
  species differences in, 136–137
  testing for pesticides, 54, 54t
  toxicity ratings and, 25–26, 26t
  uses of, 25–33, 36
  variables affecting, 22–25
Lead, 87, 142, 144, 147–149, 149, 240
Learning disorder(s). *See* Childhood behavior disorder(s)
Legionnaire's disease, 211
Lehrer, Tom, 191–192
Lethal dose. *See* LD50
Leukemia(s), 240, 241, 275
Levoglucosan, 166
Ligand(s), 143–144, 144t, 145–146
Lindane, 120, 122
Linear alkyl sulfonate(s) (LAS), *192,* 193
Lipid(s), 60–61, *61*
Lipid solubility, 62–63
Liquified petroleum gas (LPG), 42, 43t
Liver, 74–75, 79, 80, 148, 263

Liver cancer, 240
Love Canal, 234–235, 242
Lovelock hypothesis, 155–156
LPG. *See* Liquified petroleum gas (LPG)
Lung(s). *See* Respiratory tract
Lung cancer, 70, 214, 215, 240

Magnesium, 57, 194
Malaria, 30, 115
Malathion, 89t, 94, 109
Marihuana, 107
Market Basket Survey Program, 93
Maximal permitted intake (MPI), 94t, 95
Maximum tolerated dose (MTD), 278, 281
Mean lethal dose, 15–16
Measles (rubella) virus, 284
Meat Inspection Act, 86
Mechanistic toxicology, 4
MEK. *See* Methyethyl ketone (MEK)
Membrane(s). *See* Biologic membrane(s)
Mercuric mercury, 144t
Mercuric nitrate, 4
Mercury, 4, 99, 142, 145, 146
Metabolic tolerance, 268, *268*, 269–270
Metal(s). *See* Heavy metal(s)
Metasilicate(s), 201
Metastatic cancer(s), 276
Methanol, 246, 256–257, *257*, 262
Methemoglobin, 177
Methemoglobinemia, 178–179, 180
Methoxychlor, 120
Methyethyl ketone (MEK), 246
Methyl isocyanate, 109–110
Methyl mercaptan, 159–160
Methyl mercuric chloride (CH_3-HgCl), 99
Methylbenzene, 240
Methylene chloride, 211
Methylmercury, 99–100, 142, 143, 144t, 144–145, 243
Methylphenidate, 222
Microsome(s), 79
Migration of waste, 247–248
Miller Pesticide Amendment, 88
Minimal lethal dose (MLD), 15
Mirex, 45, 45t, 46, 370
Mitogenesis, 280, 281
MLD. *See* Minimal lethal dose (MLD)
Mold(s), 99, 102
Molybdenum, 142, 175–176
"Monkey children," 100
Monocyclic aromatic hydrocarbon(s), 238, 239t
Morphine, 3, 47
MPI. *See* Maximal permitted intake (MPI)
MTD. *See* Maximum tolerated dose (MTD)
Mucus, 68
Mueller, Paul, 114
Mutagenesis, *275*, 282–283
Mycoses, 98
Mycotoxin(s), 99
Myoglobin, 181

N_2O. *See* Nitrous oxide (N_2O)
N-P-K-code, 176
NA. *See* Nitrosamine(s) (NA)
Na_2S. *See* Sodium nitrite (Na_2S)
Na_2SO_3. *See* Sodium sulfite (Na_2SO_3)
NaCN. *See* Sodium cyanide (NaCN)
Nail(s), 75
Naphthalene, 68
Nasopharyngeal area, of respiratory tract, *67*, 69
National Academy of Sciences, 215
National Ambient Air Quality Standard, 212
National Center for Atmospheric Research (NCAR), 155–156, 157
National Environmental Protection Act, 84
National Institute of Environmental Health Sciences, 276
National Institute of Health, 229
National priority list (NPL) of waste sites, 237–238, 238t
Natural salicylate(s), 221
Natural salicylate radical(s), 220
Nematocides, 98
Nerve gas, 109
Neuronal membrane fluidity hypothesis, 268–269, *269*
Neuronal tolerance, 268–269, *268*
Neurotransmitter(s), 8
NH_3. *See* Ammonia gas (NH_3)
Nicotine, 108. *See also* Cigarette smoking
Nitrate(s), 176, 180–182
Nitric (HNO_3) acid(s), 167
Nitric oxide (NO), 180
Nitrilotriacetate, 201–202
Nitrite, 27–29, *27*, 32, 39–40, 177–179
Nitrogen, fatality and, 68
  fixation and, 172, *173*, 174–175
  fixed nitrogen and fertilizers, 175–176
  and life, 57, 172–173, 198
  nitrates, 180–182
  nitrite in the environment, 177–179

nitrogen cycle, *173*, 174–175
nitrogen oxide gas(es), 180
oxidation states of, 173t
oxidizing-type air pollution and, 183–184
Nitrogen dioxide (NO_2), 180, 211, 212
Nitrogen oxide(s) (NO_X), 39, 180, 183–184, 216
Nitroglycerin, 12, 71
Nitrosamine(s) (NA), 181–182
Nitrous oxide (N_2O), 175, 270
NO. *See* Nitric oxide (NO)
NO_2. *See* Nitrogen dioxide (NO_2)
NOEL (no observable effect level), 13, 37, 39, 40, 51, 52, 57, 94–95, 94t
Nonionic surfactant(s), 190, 193
Nonphosphate builder(s), 202–205, *203*, *204*
Normal distribution, 16
NO_X. *See* Nitrogen oxides (NO_X)
NPL. *See* National priority list (NPL) of waste sites
NTA. *See* Sodium nitrilotriacetate (NTA)
Nutrimaster, 134

Occupational dust(s), 67
Occupational Health and Safety Agency (OSHA), 158
Octanol-water (o/w) partition, 63, 247
Off, 107
Oil/water (o/w) partition coefficient, 63, 247
Oncogenicity testing, 277–281
OPI. *See* Organophosphate insecticide(s) (OPI)
Orfilia, Mattieu Joseph Bonaventura, 3
Organogenesis, 285
Organomercurial(s), 99
Organophosphate insecticide(s) (OPI), 109
Orthophosphate, 198
OSHA. *See* Occupational Health and Safety Agency (OSHA)
Oxidative biotransformation, 78–79
Oxygen, 57
Ozone, 42, 43t, 168, 183–184

P450 cytochrome, 136, 145, 264
P450 isozyme(s), 79, 80, 81, *81*, 108, 136, 181, 264, 269
PAH. *See* Polycyclic aromatic hydrocarbon(s) (PAH)
2-PAM, 109
Paper chromatography, 91
Paracelsus, 3
Paracelsus Principle, 3, 12
Paraquat, 106–107
Parathion, 26, 109
Particulate(s), 68–69, 160–168, 183, 211, 216
Passive diffusion, 63
PBB. *See* Polybrominated biphenyl(s) (PBB)
PCB. *See* Polychlorinated biphenyl(s) (PCB)
PCN. *See* Polychlorinated naphthalene(s) (PCN)
PCP. *See* Pentachlorophenol (PCP)
Pellizzari, Edo D., 212
Penicillin, 45, 45t
Pentachlorophenol (PCP), 100–101, 130, 215–216
Peptide(s), 145
Perchlorethylene, 216
Peroxyacetyl nitrate(s), 183
Pesticide(s). *See also* DDT
avicides, 107
banned pesticides, 86, 114
fungicides, 98–101
herbicides, 103–107
insect repellants, 107
insecticides, 107–111
resistance to, 110–111
rodenticides, 101–103
species differences and, 22
testing for, 53–56, 54t, 55t
Pesticide law(s), banned pesticides, 86
EPA and, 84–86, 88
history of, 88
"indemnification" rule, 85–86
labels, 84–85
purity of pesticides, 128
setting tolerances for, 84–86, 89t, 91–96
pH, 62–63, 72, 75, 203–204, 245–246
Phenobarbital, 80
Phenol, 42, 43t
Phenol-formaldehyde, 213
Pheromone(s), 111
pHisohex, 137
Phosgene, 110, 241
Phosphate, 198–200
Phosphate-free detergent(s), 200–202
Phospholipid(s), 60–61, *61*
Phosphorus, 57, 101, 198, 207
Photosynthesis, 105
Phytoplankton, 197–200
Pinocytosis, 64, *64*, 69
Piperonyl butoxide, 108
pK_a, 62–63
Placenta, 285–286

Pneumoconioses, 70
Point mutation(s), 282
Polarography, 91
Pollution. *See* Air pollution
Polybrominated biphenyl(s) (PBB), 134, 136t
Polychlorinated biphenyl(s) (PCB), 95–96, 131–133, *131*, 135, 136t, 241, 243
Polychlorinated naphthalene(s) (PCN), 128–130, *129*, 131, 135, 136t
Polycyclic aromatic hydrocarbon(s) (PAH), 165–166
Polymer(s), 195–196
Polyphosphate(s), 193–197, 194t, 195t, 206
Porphyria, 149
Porphyria cutanae tarda, 136
Potassium, 57
Potency ratio (PR), 19–20, 28–29, 32–33
Potentiation, 26–27, 30
Pott, Percival, 276–277
Ppm (parts per million), 38
Pregnancy, 284–286
President's Council on Environmental Quality, 202
Probit graph paper, 18–19, *18*
Procarcinogen(s), 278, 280
Proof system for expressing alcohol content, 260–261, *261*
Protein(s), 60–61, *61*, 145
Prothrombin, 102
Proto-oncogene(s), 274, 278
Pure Food and Drug Law, 86, 87
Pylorus, 262, 263
Pyrethrin(s), 108
Pyrethroid(s), 108
Pyrethrum, 108

Quantal response, 15
Quantitative toxicology, 4

Radioactive substance(s), 238
Radon, 214–215
Rall, David P., 276
Ramazzini, Bernardino, 3
Random motion, 63
Raticate, 103
RCRA. *See* Resource Conservation and Recovery Act (RCRA)
Red blood cell(s), 148–149
Red No.3, 92
Renaissance, 3
Resistance, to pesticides, 110–111
Resource Conservation and Recovery Act (RCRA), 252
Respiratory tract, absorption of aerosols through, 69, 69t
  absorption of lead through, 147
  absorption of poisonous gases through, 68–69
  absorption through, 66–70, *67*, 69t
  anatomy of, 66–68, *67*
  diseases produced by inhaled chemicals, 70
  ethanol and, 261–262
  excretion and, 74
  heavy metals and, 143
Response, all-or-none (quantal), 15
Rhizobium, 174–175
Risk assessment, 8, 244–251
Rodenticide(s), 101–103
Romans, 3
Roosevelt, Theodore, 86
Rotenone, 108
Rubella virus, 284
Ruckelshaus, William, 126, 202
Rutgers 6–12, 107

Saccharin, 92
Safety evaluation(s), 4
Safety factor (SF), 94t, 95
Salicylate(s), 220, 221
Saliva, 75
Salt, 103
Saponification, 189
SARA. *See* Superfund Amendments and Reauthorization Act (SARA)
Sedative(s), 78
Seminal fluid, 75
Sevin, 109
SF. *See* Safety factor (SF)
Short-term exposure limits (STEL), 42–43, 43t
Shoxin, 103
Sick building syndrome, 212–213
Sickle cell anemia, 282
Sign(s), versus symptom(s), 25
Silicate(s), 201
Silo fillers' disease, 180
Silvex, 86
Sinclair, Upton, 86
Single dose toxicity, 15–20
Sister chromatid(s), 280
Site description, 245–247
Skin, absorption of lead, 147
  absorption through, 73–74
  structural features of, 73–74, *73*
Small intestine, *70*, 71
Smog, 183
Smoking. *See* Cigarette smoking
SO_2. *See* Sulfur dioxide (SO_2)

Soap, historical survey of, 188–191, *189*
Socrates, 3
Sodium, 57
Sodium azide, 27–29, *27*
Sodium bicarbonate, 75, 202
Sodium carbonate, 201, 202, 203, *203*, 204
Sodium chloride, 103
Sodium cyanide (NaCN), 27–29, *27*, 45, 45t, 242
Sodium fluoroacetate (Compound 1080), 22
Sodium metasilicate, 201, 202, 203, *203*, 204
Sodium nitrilotriacetate (NTA), 201–202, 206
Sodium nitrite (Na_2S), 27–29, *27*, 39
Sodium pyrophosphate, 196
Sodium sesquicarbonate, 202
Sodium silicate, 194, 194t, 201, 202
Sodium sulfide, 27–29, *27*, 36
Sodium sulfite (Na_2SO_3), 166
Sodium tetrapropylene benzene sulfonate, 191, *192*
Sodium thiosulfate, 29
Sodium tripolyphosphate, 196, 203, *203*, 204
Species differences, in LD50, 22–23
Steinfeld, Jesse, 202, 203
Stomach, *70*, 71
"Stomach pump," 72
Stratum corneum, 73, *73*
Stratum germinativum, 73, *73*
Strychnine, 101
Subacute testing, 44
Subchronic toxicity testing, 44, 50–53, *52*
Sulfur dioxide (SO_2), 211
Sulfanilamide, Elixir of, 87–88
Sulfhydryl (-SH), 146
Sulfide(s), 27–29, *27*, 32–33, 32t, 40, 241
Sulfite(s), 166
Sulfite salt(s), 166
Sulfur, absorbed sulfites, 166
  acid rain, 167–168
  natural occurrence of, 57, 145
  oxidation of, 153, 153t
  particulates and, 160–168
  production of, 152–153
  sulfur cycle, 153–160, *154*, 156t
  woodsmoke, 165–166
Sulfur dioxide (SO_2), 39, 42, 43t, 153, 157, 160–165, *164*
Sulfuric acid (H_2SO_4), 153, 167
Sulfurous acid (H_2SO_3), 166
Superfund Amendments and Reauthorization Act (SARA), 237
Surfactant(s), 188–193, *192*, 194t
Surgeon General. *See* United States Surgeon General
Sweat, 75
Symptom(s), versus sign(s), 25
Synergism, 27, 29, 108
Synthetic reaction(s), 81

1,1,1-trichloroethane, 241
2,3,7,8-tetrachlorodibenzo-p-dioxin (TCDD), 22, 26, 80, 135–137, *135*
2,4-T, 105–106
2,4,5-T, 86, 130, 134, 135, 136t
2,4,5-trichlorophenol(s) (TCP), 130, *131*, 137
Table salt, 103
Talc, 25
Tartrazine, 5, 220, 228
TCDD. *See* 2,3,7,8-tetrachlorodibenzo-p-dioxin (TCDD)
TCE. *See* Trichloroethylene (TCE)
TCP. *See* 2,4,5-trichlorophenol(s) (TCP)
Temik, 109
Teratogenesis, 55t, 56, *275*, 283–286
Testing. *See* Chronic toxicity testing; Subchronic toxicity testing; Toxicity testing
Tetrachlorethylene, 212
Tetrachloroethylene, 241
Tetraethyllead, 143, 145
Thalidomide, 135, 284–285
Thallium, 101
Theoretical maximum residue concentration (TMRC), 94t, 95
Thin layer chromatography, 91
Thiosulfate, 33
Threshold limit values (TLV), 41–43, 43t
Time-weighted average (TWA) concentration, 42
Titanium, 142
Titratable alkalinity, 204
Titration curve(s), 204–205, *204*
TMRC. *See* Theoretical maximum residue concentration (TMRC)
Tobacco smoking. *See* Cigarette smoking
Tolerance, to ethanol, 46–47, 267–270, *268*, *271*
  in food and pesticide laws, 88–96, 89t
  to heroin, 46–47
Toluene, 68, 240, 246
Toxaphene, 120

Toxic, definition of, 12
  hazardous compared with, 12
Toxicity, acute or single dose toxicity, 15–20
  age and body weight and, 23
  ambient temperature and, 24
  chronic toxicity, 43–47
  diets and, 23–24
  health status and, 23
  interactions between two chemicals in body, 26–32
  relationship of dose to, 14
  route of chemical exposure and, 24
  sex differences in, 23
  signs of, 25
  species differences in, 22–23
  subacute testing, 44
  subchronic testing, 44
  tests on laboratory animals, 13
Toxicity rating(s), 25–26, 26t
Toxicity testing. *See also* Chronic toxicity testing; Subchronic toxicity testing
  acute toxicity testing, 53–56, 54t, 55t
  chronic toxicity testing, 50, 55t, 56
  EPA requirements for, 53–56
  FDA requirements for, 53
  subchronic toxicity testing, 50–53, *52*
  trace elements and, 56–57, *57*
Toxicological evaluation, 249
Toxicology, goals of, 12–13
  historic insights on, 2–8
  inhalation toxicology, 36–47
  mechanistic toxicology, 4
  quantitative toxicology, 4
Trace elements, 56–57, *57*
Train, Russel, 202
Transport processes, 61–66, *62, 64, 65*
Trazine herbicide(s), 105
Trichloroethylene (TCE), 212, 239–240
Tropism(s), 103
Tube worm(s), 158–159
Tumor(s), 274–275, 277–279
TWA concentration. *See* Time-weighted average (TWA) concentration
Tylenol, 25
Typhus, 101, 114

UCR. *See* Unit cancer risk (UCR)
Unit cancer risk (UCR), 250–251
United States Public Health Service, 88, 117
United States Surgeon General, 201–202, 215
Urine, 66, 75, 148, 149, 150, 263
USPHS. *See* United States Public Health Service

Valium, 269
Vicks Vaporub, 107
Villi, *70*, 71
Vitamin A, 128
Vitamin K, 103

Warfarin, 45, 45t, 46, 53, 102–103
Waste site(s). *See* Hazardous waste site(s)
Water contamination. *See* Groundwater contamination
Well water methemoglobinemia, 178–179
Went, Fritz, 104
Wine, 260–261
Woodsmoke, 165–166

Xenobiotics, 78–79, 80, 283
Xylene, 246

Yellow fever, 115
Yusho disease, 133, 136

Zero-phosphate detergent(s), 206
Zero tolerance, 90–91
Zinc, 142, 149
Zinc ammonium sulfate, 163–164, *164*